'There is a gap in knowledge in this area and this book will help students and practitioners to better understand the financial implications of their design decisions.'

Paul Falconer, Chairman, Falconer Chester Hall Architects

'The contents of this book will provide architects and urban designers with a more rounded view of finance and its importance in sustainable design. It would be a useful addition to the industry.'

Professor Greg Beattie, Associate Director, Arup

'At Urban Splash we are striving to create zero carbon buildings. They clearly have to be economically viable; this book will be of great interest to us and a very useful addition to the academic and industrial literature.'

Simon Humphreys, Development Director, Urban Splash

New Financial Strategies for Sustainable Buildings

Built environment professionals considering whether to embark on the design and construction or retrofit of a fully 'sustainable' or 'green' build need to know the financial implications of their decisions. What are their financial options? What are the risks? This book offers practical guidance on how sustainable building projects are financed, designed and built. All too often sustainable building is undertaken without proper consideration of the true lifecycle cost, risk and financial impact. This book will take the reader on a journey from initial sustainable design through to final completion highlighting the finance options available to them.

New Financial Strategies for Sustainable Buildings provides key guidance to a variety of professionals, including architects, designers, contractors, construction managers, investors and other interested parties, while providing a useful reference to students on architecture, construction management and real estate/surveying courses who need to know about finance, construction economics, and sustainable development projects.

Stephen Finnegan is a low-carbon design specialist and lecturer in sustainable architecture at the University of Liverpool, UK.

New Financial Strategies for Sustainable Buildings

Practical Guidance for Built Environment Professionals

Stephen Finnegan

LONDON AND NEW YORK

First published 2018
by Routledge
2 Park Square, Milton Park, Abingdon, Oxon OX14 4RN

and by Routledge
711 Third Avenue, New York, NY 10017

Routledge is an imprint of the Taylor & Francis Group, an informa business

British Library Cataloguing-in-Publication Data
A catalogue record for this book is available from the British Library

Library of Congress Cataloging-in-Publication Data
Names: Finnegan, Stephen, author.
Title: New financial strategies for sustainable buildings : practical guidance for built environment professionals / Stephen Finnegan.
Description: Abingdon, Oxon ; New York, NY : Routledge, 2018. | Includes bibliographical references and index.
Identifiers: LCCN 2017005844|
ISBN 9781138680265 (hardback : alk. paper) |
ISBN 9781138068520 (pbk. : alk. paper) | ISBN 9781315563725 (ebook)
Subjects: LCSH: Sustainable buildings. | Sustainable buildings—Design and construction—Costs. | Real estate development—Finance.
Classification: LCC TH880 .F535 2018 | DDC 690.028/6—dc23
LC record available at https://lccn.loc.gov/2017005844

ISBN: 978-1-138-68026-5 (hbk)
ISBN: 978-1-138-06852-0 (pbk)
ISBN: 978-1-315-56372-5 (ebk)

Typeset in Baskerville
by Florence Production Ltd, Stoodleigh, Devon, UK

Printed in the United Kingdom
by Henry Ling Limited

Contents

Figures

Tables

Preface

This practical guidance addresses the two fundamental, but often disparate, issues of sustainable design and finance in buildings. For many years sustainability and in particular sustainable buildings have been seen as a 'desirable' but costly venture. Creating a new sustainable building or retrofitting an existing building with new low carbon technical solutions can be an expensive venture; with return on investment not always apparent from the onset. However, this paradox is changing and new sustainable and cost-effective opportunities and solutions exist. Moreover, there are a large amount of emerging case studies that support the idea that sustainability and cost reduction can coexist. It is now possible to create a sustainable building in a cost-effective manner but only if the participants are willing to embrace change.

It is crucially important for built environment professionals i.e. architects, construction and project managers, surveyors, engineers etc. to be aware of the conventional and new options available so that this 'desirable' ideal is quashed. Through the lifecycle of a building, the professionals involved should all have a basic understanding of the financial implications, opportunities and consequences of choice. When an architect envisions a new building design and includes for example Solar Photovoltaics, a new Ground Source Heating System, Wind Turbines or super insulation; they should have an idea of how these technologies could be funded. They should also stop to think why? Should they not be considering an enhanced building fabric? A better building orientation? A new approach to managing energy demand? Thought should also be given to how realistic these options are, have they been field tested and what are the resultant financial implications. More specifically, the architect should consider the impact the design choice will have to the build cost. Is it financially viable? What about the lifecycle cost? Similarly, when a construction or project manager, surveyor or engineer specifies how the building will be built, they too should be asking the same questions and fully appreciated the consequences of their choices. This is why this practical guide is necessary and relevant. Throughout this book there are a series of guiding principles to assist the reader in making the right decisions at the right stage in the development. These guiding principles are the crucial elements for success.

The opening Chapter 1 considers the issue of financing sustainable buildings, investigating the modern role of the built environment professional in sustainable design. This is followed by an overview of sustainable building design (both passive and active) and finance in principle and practice. This chapter concludes with a review of financing the latest modern sustainable buildings; highlighting lessons learnt and key considerations.

Chapter 2 investigates the traditional and new methods of financing sustainable buildings. Reviewing common approaches, government schemes, self and third-party financing. The use of Energy Performance Contracts (EPCs) and Energy Services Companies (ESCOs) is also discussed in detail with case studies highlighting the effectiveness of these types of schemes.

Incentives to investment and the reasons for producing a sustainable building are discussed in Chapter 3. Cost savings, increased building value, occupier productivity and commitment to sustainability are all explored. These incentives highlight the opportunities available and modern necessity for sustainable buildings.

In addition to incentives, there are also barriers and Chapter 4 details the common barriers to investment. In this chapter a discussion on capital expenditure, split incentives, interchangeable support from governments, new technology and commitment to change are researched.

The penultimate Chapter 5 considers the main obstruction to the wholesale uptake of sustainable buildings, that of risk and change in procedure. Financial and technical risk are clear hurdles to overcome, along with making short- vs long-term investment decision and considering the ever-changing market conditions, standards and requirements.

A series of successful ESCO funded case studies are then presented in Chapter 6 highlighting the possibilities and opportunities for a mixture of sustainable building types. You may find solace in these case studies and consider the effectiveness for your own building design. You may conversely consider the options involved to be profligate and inconsistent. The hope is that readers of this book will make their own decisions, reflect on the possible, manage risk through acquiring new knowledge and talk the universal language of cost saving through better environmental design. Only then will sustainable buildings become a global reality.

'You never change things by fighting the existing reality. To change something, build a new model that makes the existing model obsolete.' – Buckminster Fuller

Dr Stephen Finnegan
Lecturer in Sustainable Architecture, University of Liverpool
December 2016

List of abbreviations

AGDIIS	Australian Government Department of Industry, Innovation and Science
AIA	American Institute of Architects
ARPT	Autorité de Régulation de la Poste et des Télécommunications
BATNEEC	Best Available Technology Not Entailing Excessive Cost
BBP	Better Building Partnership
BedZED	Beddington Zero Emission Development
BEMS	Building Energy Management System
BIM	Building Information Modelling
BIQ	Bio Intelligent Quotient
BMS	Building Management System
BRE	Building Research Establishment
BREEAM	Building Research Establishment Environmental Assessment Method
CAPEX	Capital Expenditure
CASBEE	Comprehensive Assessment System for Built Environment Efficiency
CBA	Cost Benefit Analysis
CBD	Commercial Building Disclosure
CCL	Climate Change Levy
CHP	Combined Heat and Power
CIBSE	Chartered Institute of Building Services Engineers
CIOB	Chartered Institute of Building
CIT	Corporate Income Tax
CPR	Construction Products Regulation
DEC	Display Energy Certification
DECC	Department of Energy and Climate Change
EBRD	European Bank for Reconstruction and Development
EC	European Commission
ECA	Enhanced Capital Allowances
ECO	Energy Company Obligation
EEF	Energy Efficiency Financing
EEM	Energy Efficiency Measures

EEVM	Energy Efficiency Value Matrix
EIB	European Investment Bank
EPA	Environmental Protection Agency
EPBD	Energy Performance Building Directive
EPC	Energy Performance Contract
EPCs	Energy Performance Certificates
EPD	Environmental Product Declaration
ESCO	Energy Services Company
ETL	Energy Technology List
FBC	Full Business Case
FEES	Fabric Energy Efficiency Standard
FIT	Feed-in-Tariff
FM	Facilities Management
GBA	Green Building Alliance
GBI	Generation Based Incentives
GBS	Government Buying Standards
GCS	Green Certification System
GIB	Green Investment Bank
GSA	General Services Administration
GSHP	Ground Source Heat Pump
GTFS	Green Technology Financing Scheme
HEFCE	Higher Education Funding Council of England
HVAC	Heating, Ventilation and Air Conditioning
IBE	Institute for Building Efficiency
IECC	International Energy Conservation Code
IGP	Investment Grade Proposals
IPD	Integrated Project Delivery
LCA	Life Cycle Assessment
LCCA	Life Cycle Costing Analysis
LED	Light Emitting Diode
LEED	Leadership in Energy and Environmental Design
LPG	Liquified Petroleum Gas
LZC	Low and Zero Carbon
M&E	Mechanical and Electrical
M&V	Measurement and Verification
MMC	Modern Methods of Construction
MOHURD	Ministry of Housing and Urban–Rural Development
MRV	Monitoring, Reporting and Verification
MVHR	Mechanical Ventilation and Heat Recovery
NPV	Net Present Value
NZEB	Nearly Zero-Energy Buildings
OBC	Outline Business Case
OFGEM	Office of Gas and Electricity Markets
OJEU	Official Journal of the European Union
OPEX	Operational Expenditure

PCR	Public Contracts Regulations
PFI	Private Finance Initiative
POE	Post-Occupancy Evaluation
PPA	Power Purchase Agreement
PPP	Public Private Partnership
PrOACT	Problem, Objective, Alternatives, Consequences and Tradeoffs
PROBE	Post-Occupancy Review of Buildings and their Engineering
PV	Photovoltaic
QA	Quality Assured
QC	Quality Control
QM	Quality Management
RGF	Revolving Green Fund
RHI	Renewable Heat Incentive
RO	Renewables Obligation
ROCs	Renewables Obligation Certificates
ROI	Return on Investment
RICS	Royal Institute of Chartered Surveyors
SBS	Sick Building Syndrome
SCI	Sustainable Construction and Innovation Group
SEELS	Salix Energy Efficient Loans Scheme
SFS	Siemens Financial Services
SIP	Structured Insulated Panels
SOP	Strategic Outline Case
SWOT	Strengths, Weaknesses, Opportunities and Threats
TFT	Tuffin Ferraby Taylor
UNEP	United Nations Environmental Programme
UKGBC	UK Green Building Council
USEPA	US Environmental Protection Act
USGBC	US Green Building Council
VAT	Value Added Tax
VfM	Value for Money
VM	Value Management
WHD	Warm Home Discount
WGBC	World Green Building Council

Guiding principles

Throughout this book there are a series of guiding principles to assist the reader in making the right decisions at the right stage in the development of a sustainable building. These guiding principles are the crucial elements for success and are listed at the end of each chapter. The full list is presented below.

Principle 1.1 – Cost effective sustainable buildings can and are being built but there is an additional capital expenditure.

Principle 1.2 – For all developments, the client must commit to producing a sustainable building and in doing so they commit to a higher capital expenditure for longer-term gain.

Principle 1.3 – Producing a sustainable building in principle is simple. Take a 'fabric first approach' and then consider passive and active design.

Principle 1.4 – Producing a sustainable building in practice is complex. Most building owners still consider sustainable design as a 'nice to have' feature that is too capital expenditure intensive. There are multiple and conflicting beneficiaries and a general unwillingness to invest upfront.

Principle 2.1 – Self- financing energy efficiency measures is the best option if the monies are available and a team of internal experts exist who are not concerned by the potential changes in procedure and risk. Government schemes can be useful to assist in financing energy efficiency measures however they should not be a major contributing factor.

Principle 2.2 – An important principle of an Energy Performance Contract (EPC) is that energy efficiency investments are paid for over time by the value of energy savings achieved.

Principle 2.3 – The use of an Energy Services Company (ESCO) is key in achieving a cost neutral zero carbon development for those who do not have the available capital and as concerned over the perceived risk.

Principle 3.1 – A sustainable building can be more valuable than a standard equivalent building and the scale of that value is dependent upon the method with which the energy efficiency measures were funded in the design stage.

Principle 3.2 – Generally speaking a sustainable building improves health and well-being through productivity of occupants and users of the building, reduced emissions and use of natural resources.

Principle 3.3 – A sustainable building ensures that current and future regulations are met and that the building remains attractive to its owners and occupants over a longer time period. Thereby increasing the return on investment.

Principle 4.1 – Sustainable buildings require a commitment to change and from a cost perspective should be considered using Life Cycle Costing Analysis (LCCA) and not capital expenditure only.

Principle 4.2 – The incentive to build a sustainable building is dependent upon who the beneficiary is. Sometimes the incentives are split and the relationship between the owner and occupier is crucial.

Principle 4.3 – The current incentives and standards are ineffective in encouraging the widespread introduction of sustainable buildings. There is a growing need to develop simple but comprehensive assessment tools.

Principle 4.4 – Installing new energy efficiency measures (EEMs) will require a basic understanding of existing building services and the provision of new procurement and tendering procedures.

Principle 4.5 – Changing government policy and current building regulations are acting as a barrier to the uptake of sustainable buildings.

Principle 5.1 – Financial risk can be eliminated through the involvement of a third-party ESCO and the introduction of an EPC.

Principle 5.2 – Technical risk can be reduced significantly by careful planning, use of established techniques and a willing to change processes.

Principle 5.3 – Sustainable buildings should be considered in the design stage and choosing the right combination of energy saving options is crucial to success.

Principle 5.4 – Making investment decisions is subjective and dependent upon your personal circumstances. They must be considered on a long-term basis.

1 Financing sustainable buildings

It is recognised that buildings and their operation contribute to a large percentage of total energy end-use worldwide. Buildings account for one-third of global greenhouse gases with commercial and residential buildings alone accounting for 40 per cent of the world's energy consumption, RICS (2015), UNEP (2007). The vast majority of energy is consumed by existing buildings while the replacement rate of existing buildings by new-build is only around 1 to 3 per cent per annum, Barlow and Fiala (2007), Roberts (2008).

Clearly then there is a need for more sustainable buildings if we are to meet our global carbon reduction targets and indeed one should start by assessing the existing stock of buildings. But what is a sustainable building? If a commercial building is fitted with solar photvoltaic (PV) panels or a green roof, does it become a sustainable building? Many use rating systems to explain what a sustainable building is i.e. the US Leadership in Energy and Environmental Design, LEED (2016), the UK Building Research Establishment Environmental Assessment Method, BREEAM (2016), the Australian Green Star Programme, GBCA (2016), the Japanese Comprehensive Assessment System for Built Environment Efficiency, CASBEE (2016) or the Canadian Green Globes (2016). Others refer to the energy use in the building, the triple bottom line 'social, environmental and financial', operational and embodied carbon or the 'brundtland statement'.

> Sustainable development is the kind of development that meets the needs of the present without compromising the ability of future generations to meet their own needs.
>
> UN Bruntland Report 1987

The Australian Sustainable Built Environment Council (ASBEC) has created a table that explains the terminology and provides a definition by region of what is meant by a sustainable building, see Appendix 1. In this book the exact definition or building type is not important, but what is important is that all 'sustainable buildings', whether they be zero carbon, nearly zero energy, net zero or carbon neutral, can be produced in a *cost-effective manner*. The preconceived view of many is that this is simply not true and it is generally too costly to produce a new sustainable building or retrofit an existing building to become more sustainable. The World Green Building Council (2013) examined the business case for green buildings by looked at hundreds of examples. In two particular areas (design and construction and operating cost) they concluded the following:

> **Design and construction costs** – Research shows that building green does not necessarily need to cost more, particularly when cost strategies, program management and environmental strategies are integrated into the development process right from the start. While there can be additional costs associated with building green as compared to a conventional building, the cost premium is typically not as high as is perceived by the development industry.
>
> **Operating costs** – Green buildings have been shown to save money through reduced energy and water use and lower long-term operations and maintenance costs. Energy savings in green buildings typically exceed any design and construction cost premiums within a reasonable payback period. In order to achieve their predicted performance, high-performing green buildings need to be backed up by robust commissioning, effective management, and collaboration between owners and occupiers.

The element of perception is key in what is now referred to as the 'perception gap' – *the estimated vs. actual cost premiums for green buildings*. The actual cost premium of a green building over a conventional building, based on the numerous studies referenced in the World Green Building Council report is –0.4 to 12.5 per cent. The perceived cost is 0.9 to 29 per cent.

This book supports the claims made by the World Green Building Council in that financing both existing and new sustainable buildings is not a costly venture as a number of new financial strategies now exist. In the modern built environment there are vast opportunities to create sustainable cost-effective buildings and if the client and project team are willing to consider the alternatives and embrace change then this can be achieved. Those individuals, who take the time to understanding the processes, learn from successful case studies and take calculated risks, are finding that sustainable buildings pay. Edwards and Naboni (2013) found that in the United States and Europe researchers have now discovered that buildings based on more ecological approaches, i.e. green sustainable buildings led to a social and economic benefit for the developer.

It is widely acknowledged that the fear of higher investment costs, in comparison to a conventional building, is the main barrier to the production of new sustainable building, Häkkinen and Belloni (2011), Hydes and Creech (2000); Larsson and Clark (2000); Nelms, Russel, and Lence (2005). Before one can understand how sustainable buildings can be financed, it is first of all necessary to understand how they are built and what they are. The starting point is to appreciate the modern roles of built environment professionals.

The modern role of the built environment professional

Built environment professionals are involved in each stage of the life of a building, from concept design through to construction, operation and final disposal or reuse. Typically, there are two main types of construction, each with different contractual arrangements. It is very important to understand these types of arrangements as they form the basis upon which all buildings are constructed. As a result, when a sustainable building (with a new financial strategy) is considered, a new arrangement with a contractor and/or consultant is necessary – this is discussed in later chapters. This is the first 'sticking point' when considering sustainable buildings, as it requires a change to the conventional approach.
The two main types of construction are as follows:

1. Design Bid Build – whereby the client would appoint consultants to design the project and prepare tenders. Contractors then competitively tender for the work and then have responsibility for the project and;
2. Design and Build – whereby one entity works under a single contract. Typically, the contractor is appointed directly to design and construct the works.

With both construction types there is a need for specialist services that will bring the relative parties together to create the building. This would include for example:

- The geotechnical engineer who would examine and explore the site for ground conditions and determine the design for foundations and earthworks.
- A land surveyor who would then determine boundaries and relative positions for construction.
- The architect who creates the vision and ensures that contractors follow their design and plan of work; use the correct materials and ensure quality management.
- The structural engineer who would ensure that structural calculations are correct, that the building will remain upright and that all loads and forces are strong enough to avoid collapse.
- The quantity surveyor, who is concerned with the value of construction works is involved from early design and in production of the essential bill of quantities.
- The building services engineer who would interact with the teams to ensure services are correctly sized and installed to their specification. They would also handle Mechanical and Electrical (M&E) works, design layouts and ensure safety of systems.

Finally, a mixture of construction, facility and project managers, surveyors and engineers are involved at various stages in the planning, construction and operation of the premises. All of which have a major role to play in influencing and implementing change. When considering how to finance a sustainable building it is therefore important to understand how a new strategy could impact on each of these individual parties and what are the constraints.

Considerations and constraints

Each built environment professional should, as part of their role, consider their involvement in the production of a sustainable building. Clearly there needs to be an overarching commitment from the client and contracting body to achieve this goal. In most modern sustainable buildings this commitment is in place from the outset and the next key step is to actively involve all participants. There is no requirement for each professional to have expert knowledge and in-depth experience of the entire process; however, each should understand and appreciate their role in providing advice and their constraints in making decisions that will impact on the overall production of a sustainable building. The key questions are: what role should each person play? what constraints will they face? and what should they as individuals be doing? This is clearly a complex and difficult set of questions to answer; nonetheless some guidelines are presented in Table 1.1.

These guidelines may work well in principle but what about the practicalities? How do the professionals work in collaboration? How do we successfully integrate the teams? Before we can answer these questions, we need to understand the principles and practicalities.

1.1 Sustainable building design and finance in principle

Ask any built environment professional about the financial implications of a sustainable building and you will likely get the same response '*It is too costly and too risky*'. In the modern day, this statement is wholly unjustified and as discussed earlier is a perceived risk. Edwards and Naboni (2013) state that buildings based on more ecological approaches lead to social and economic benefits for the developer. In more detail these benefits are grouped under four headings:

1 better lifecycle costing;
2 improved productivity or performance in functional terms;
3 better social relationships at a building and community level;
4 enhanced image for the building and the organisation responsible for its inception.

All of which may well be true but when a developer is faced with an option of retrofitting an existing building or considering a new build; sustainability features are still a high initial capital expenditure and as such are generally quickly discounted. This is because most decisions on major capital works are made on a short-term basis. All too often, longer-term *lifecycle costing* is given little consideration. Bull (2003) states that the 'lowest-cost' method of decision making is without question, the current major method of building option selection and works on the assumption that the cheapest solution is the best financial option. As maintenance,

Table 1.1 Roles of built environment professionals

Built environment professional	*Guidelines*
Building owner/client	The owner/client should commit to the design of a sustainable building and most importantly should provide a realistic vision.
Landscape architect and site planner	The landscape architect and site planner have a duty of care and legal commitment to provide the specification set. They along with other professionals need a clear vision and scope of works.
Architect	Clearly the architect as master designer has a key role to play in the vision and design of the building. As a result they should consider the fabric first approach, passive and active design and the use of low and zero carbon technology (if necessary).
Acoustic consultant	The acoustician should consider any impact on acoustics should new energy efficiency measures be introduced that could alter the shape and form of the building i.e. will a new passive cross ventilation system alter the acoustics of a building?
Geotechnical engineer	At the planning stage there may be a requirement for a geotechnical engineer to consider the impact of the building on the subsurface. For example if a new building requires piles could they be lagged with pipes for a new ground source heating or cooling system?
Structural engineer	The structural engineer will need to understand the proposed measures for the sustainable building and ensure that they are both safe and viable. For example can an existing roof support the weight of a new solar photovoltaic array?
Interior designer	The interior designer should consider the look and feel of a sustainable building. How could they integrate the features into the building?
Electrical engineer	A new sustainable building will, in the majority of cases, require a new, amended or altered electrical system which the engineer may not be familiar with. They should consider how they will work in collaboration with other specialist services.
Construction contractor and inspector	Is the construction contractor and inspector familiar with the new processes involved in creating a sustainable building? Are they qualified to act as inspectors for unfamiliar kit and processes?
Quantity surveyor (QS)	The QS has a key role to play in costing and programme planning of new options that may be unfamiliar. Consideration of this in the bill of quantities and how it could impact other elements of the build is necessary.
Mechanical engineer	The mechanical engineer, as with the electrical engineer, may need to work with a new, amended or altered mechanical system which they may not be familiar with. They should consider how they will work in collaboration with other specialist services.
Construction manager	The construction manager will need to fully understand the stages involved and the changes to standard procedures necessary to create a sustainable building. Ultimately they will be required to manage the process and build.
Building users	On completion the building users have a crucial role to play in the maintenance and occupation of a sustainable building. They should be educated on the buildings operation and controls.

energy, management and operation costs increased, many building owners then discovered that the 'lowest-cost' system was not always the cheapest solution if they took into account the lifetime use of the building. It is clear then that one should consider the longer term. Not only will this save us money but it will make the building more sustainable.

So what is a sustainable building and how, in principle, should it be designed and financed? We start by taking a 'fabric first' approach.

1.1.1 Fabric first approach

The fabric of a building is any structure, surface, fixture or fitting associated internally or externally with a building. From a sustainable design perspective, the building fabric is the most significant component. This is why the term 'fabric first' is used within the industry. This refers to achieving a well-built, thermally efficient and air tight building and can be applied to both new build and retrofits. There are additional benefits to taking a fabric first approach and they are to lessen the need for extensive use of so called active design measures and/or new technology, reduction in energy use and therefore cost and longevity of the building. Fabric first is synonymous with the Passivhaus (2016) energy performance standard. The six key principles of which, as outlined in Hopfe and McLeod (2015), are:

1. superinsulation;
2. airtight building envelope;
3. thermal bridge-free construction;
4. compact form;
5. optimal use of passive solar gain;
6. mechanical ventilation with heat recovery (MVHR).

These key areas can be used as inspiration and guidance for a fabric first approach, while maintaining an element of design flexibility. Fabric first is a guide that does not require Passivhaus standard or certification, though if it did this may have an enduring positive impact on global building stock.

Financing a fabric first approach is reliant on the client and as such can be an obstacle due to the perception of increased capital expenditure or an opportunity if the finances are considered on a lifecycle basis. A building which takes the fabric first approach to the extreme i.e. a Passivhaus building can be funded through schemes such as EuroPHit (2016) which assists in financing energy retrofits.

1.1.2 Passive design

The principle of passive design is an approach to building design that uses the buildings architecture to minimise energy consumption and improve thermal comfort through carefully considering the performance of building elements and optimising them for interaction with the local micro-climate, Lechner (2014). Passive design looks to exclude all mechanical systems that are reliant on fossil fuels-based energy and ensure a consistent level of internal comfort for the buildings occupants. Any passive design building will require a multidisciplinary approach to problem solving and working in collaboration. It involves the development of a specialist understanding of each individual element in the construction process. Developing new financial strategies for sustainable buildings requires the same mindset. Passive design requires a consideration of the following, with explanations provided by Housladen (2012) and reproduced in Appendix 2:

(a) building use;
(b) macro-climate;
(c) micro-climate;
(d) building orientation;
(e) building shape;
(f) roof shape and angle;
(g) volume to surface area ratio;
(h) site zoning;
(i) typology;
(j) thermal;
(k) ventilation;
(l) light;
(m) thermal mass.

All of these have a role to play in visualising the level of intervention required and the financial strategy for consideration. Financing passive design can be very straightforward if the client is self-financing the project and they are comfortable with the approach. It becomes more complex when borrowing is required for new nonstandard bespoke designs which will initially cost more money. It should be noted that passive design can enable cost-effective solutions to heating and cooling of buildings if designed correctly. For example, cross ventilation can be achieved by simply opening windows on either side of a building and allowing for the direct passage of air through the building. The windows could be manually or automatically (through active design) operated depending upon internal and external climatic conditions.

The Technische Universtät Darmstadt (TUD), see Figure 1.1, provides an example of passive and active design through the creation of a solar-powered house. This highly insulated home uses automated passive louvre window shades to reduce unwanted heat-gain. The main focus of the building was the fabric and in particular the use of a façade covered in solar cells, with vacuum insulated panels and phase-change material, Dezeen (2009).

1.1.3 Active design

Active design involves the general use of fuel in a mechanical system to heat or cool a building, whereas passive design maximises the use of 'natural' sources of heating, cooling and ventilation to create comfortable conditions inside buildings.

Figure 1.1 Technische Universtät Darmstadt (TUD) passive and active design

One of the most famous UK examples of passive and active design which is used in sustainable buildings, is that of the Beddington Zero Emission Development (BedZED), see Figure 1.2, a large mixed use development located in the London Borough of Sutton. The scheme was designed with a mixture of passive design (natural wind-driven ventilation) and active design (biomass Combined Heat and Power [CHP] plant) with numerous other features. The total development costs for BedZED sum up to €17 million. The costs turned out to be 30 per cent higher than expected and the price of a BedZED home was 20 per cent higher than the average price of an apartment in the same area, BedZED (2016).

Clearly, active design should only be considered when all passive design elements have been exhausted. In reality, a number of developers and building owners will jump straight to active design prior to consideration of passive design. If the word 'sustainability' is mentioned to a building owner, they will more than likely think about a technical solution. But what about the benefits that could be gained through better use of the existing or new building? Why not first of all consider the passive features of orientation, climatic factors, shape, zoning, typology, thermal properties and most importantly operational use. Maybe there are much higher cost and energy savings by simply using less electricity, avoiding heat loss by making the building more airtight.

Active design involves the use of for example: Heating, Ventilation and Air Conditioning (HVAC) systems to actively control internal temperatures; artificial lighting in buildings where natural light cannot penetrate or is insufficient; controls such as Building Management Systems (BMS) and more advanced Building

Figure 1.2 BedZED

Energy Management Systems (BEMS). The principal role of which is to regulate and monitor HVAC and lighting control.

Financing active design is straightforward as a number of the common systems have been in use for a number of years, therefore the risk involved is minimal. In addition, a number of active design systems are 'off the shelf' products such as HVAC and LED lighting. Self-financing or borrowing for these systems, with relatively accurate payback estimates is not difficult. In addition, there are other avenues to explore such as energy efficient financing (discussed in Section 1.1.5), grants, tariffs and tax allowances.

- Grants can provide financial support for the use of particular energy efficient system.
- Tariffs such as Feed in Tariff (FIT) provide an ongoing payment per kWh of electricity generated and/or exported to the network over a fixed period of time.
- Tax allowances such as Enhanced Capital Allowances, ECA (2016), offer the opportunity to provide a more tax efficient way of purchasing new energy efficient equipment.

1.1.4 Passive and active design

Passive and active design in reality work hand in hand. There are very few buildings that use just one design approach. It is common to think of 'active design' being the opposite of 'passive design' but the strategies are not at odds with one another. Active design follows passive design techniques but takes them to the next level by producing their own energy. There are a vast amount of case studies that demonstrate the successful integration of both design approaches and all have been financed via consideration of the lifetime impact and not the initial capital expenditure. Some of the most common case studies include for example the Autorité de Régulation de la Poste et des Télécommunications (ARPT) Headquarters by Mario Cucinella Architects in Algeria. For this building the natural shape of a sand dune is used to allow for natural cooling to divert hot winds at midday and capture the cool breezes at night. In addition, a Solar PV skin is used to capture sunlight on the southern slope. A second example is the Hive by Feilden Clegg Bradley, see Figure 1.3. This is a mixed use development situated in Worcester (UK). The building is cooled using water from the nearby River Seven, has a biomass boiler which uses locally sourced woodchip to generate heat, uses chilled beams to assist cooling, embraces radiant cooling through the use of plastic pipework coils embedded in concrete slabs and is mechanically ventilated.

A third example is the GSW Headquarters in Berlin by Sauerbruch Hutton, see Figure 1.4. This building uses passive and active design to create a low energy building. For example, natural cross ventilation can be enabled when windows on both sides of the building are opened. Stack ventilation is used to remove hot

Figure 1.3 The Hive by Feilden Clegg Bradley

Figure 1.4 GSW Headquarters

air from the building. In addition, an active advanced lighting control system is used throughout the building.

All three examples above use elements of passive and active design and the financial considerations were explored at the design stage of each project. Ensuring that energy efficiency measures (EEMs) worked on a technical and financial basis.

1.1.5 Energy efficiency measures

The choice of EEM, which includes low and zero carbon (LZC) technologies, is fundamental to the production of a cost-effective sustainable building. Only when all passive design measures have been considered should one then start to consider the use of these active EEMs. There are some common technologies in use on many buildings and other options which remain possibilities until the financial payback can be realised. Table 1.2 provides a list of the possible EEMs available to the market at the time of writing. The market for new advancements in EEMs is fast moving and within 12 months it is expected that this list will feature new options.

Those measures highlighted in bold are the most common types used within the industry. The reason for their success is two-fold: (1) they are low technical risk decisions and (2) are financially viable. The remainder do not fit this profile and as such, in the majority of cases, have limited uptake. The minority cases are those of major commercial projects that are able to invest in longer-term options due to the high expensive running costs of the premises. An additional key aspect on the choice of EEM is how sustainable the product is. Choosing the right EEM

Table 1.2 List of energy efficiency measures (EEMs)

Common energy efficiency measures	*Description*
Fabric solutions	– **Cavity wall insulation** – Draught proofing – **Energy efficient glazing** – **External wall insulation** – High thermal performance external doors – Internal wall insulation – **Loft or rafter insulation** – Flat roof insulation (Warm deck – Cold deck systems or Inverted flat roofs) – Under-floor insulation – Heating pipe insulation
Low and zero carbon solutions *Heating, ventilation and air conditioning*	– **Condensing boiler systems** – Heating controls (for wet central heating system and warm air systems) – Under-floor heating systems – Heat recovery systems – Mechanical ventilation systems (predominately non-domestic use) – Flue gas heat recovery devices – High efficiency replacement warm-air units – Fan- assisted replacement storage heaters
Lighting solutions	– **LED lighting** – Effective lighting controls (control gear: ballasts)
Micro-generation technologies	– Ground and air source heat pumps – Solar thermal – **Solar photovoltiacs (PV)** – Biomass boilers – Micro-CHP – Micro-wind generation – Micro-hydro systems

should not solely be a financial decision, but should also consider the lifecycle impact. This covers both the operational and embodied energy use of the product.

1.1.6 Embodied carbon

Choosing the right material and/or technology for the building is of crucial importance. A super insulated sustainable building created using an unsustainable material is clearly at odds. This is why it is also important to consider the *embodied energy* of the proposed measure(s). It may be initially more costly to procure a more sustainable material or technology and this should be factored into any decision-making process. Inevitably the cheapest option is not always the most sustainable.

The embodied energy of materials or products is another key determinate in the design of a sustainable building. This has a direct correlation to the cost as one would expect that an 'off the shelf' material is cheaper than a bespoke sustainable material. In an ideal situation all materials and systems specified should only be chosen if they are 'sustainable' and cost-effective. Indeed, in order to achieve a BREEAM excellent rating the choice of material is significant. Embodied energy is a term that has been in use for a number of years and essentially it is the amount of energy required during the design, manufacture, transport and disposal/reuse of a product. The BRE Green Guide to Specification, BRE (2016) is a useful tool as designers are able to compare and contrast the embodied energy of different materials for use in construction. In addition to this the Environmental Product Declaration (EPD) is a verified and registered document that communicates transparent and comparable information about the lifecycle environmental impact of a product, EPD (2016). It is now considered by the European Commission (EC) as an adequate instrument to communicate the environmental performance of building products and to promote sustainable construction. As such the Construction Products Regulation (CPR) requires all construction product manufacturers to declare the environmental impact entailed in the making of their products and for that purpose they must use the EPD, CPR (2013).

It is also a key recommendation of the UK Low Carbon Construction Innovation and Growth Team in their Final Report to the Government, published in 2010, that as soon as a sufficiently rigorous assessment system is in place, the Treasury should introduce into the Green Book a requirement to conduct a whole-life (embodied + operational) carbon appraisal and that this is factored into feasibility studies on the basis of a realistic price for carbon. The idea that embodied energy calculations should be undertaken when specifying materials to be used in construction is gaining momentum, HM Government (2010).

1.2 Sustainable building design and finance in practice

The principles of sustainable building design and finance have been explained above and are clear. They are to understand the building and its use, apply a fabric first approach, consider passive and active design (with a focus on passive design first), carefully select the right EEMs and consider the materials being used. The reality for modern-day building construction is that sustainable design remains a '*nice to have*' feature. Whether it is the retrofitting of an existing building or the construction of a new building, the owner will, in the majority of cases, opt for the most cost-effective low-risk option. This could be due to the fact that most building owners are not fully aware of the new financial strategies that exist; they are unwilling to take the risk of not following a standardised procedure and/or the perceived return on investment is inadequate. This statement is supported by Pinkse and Dommisse (2009) who also commented on the need for additional outsourcing for specialist contractors and increased complexity in the maintenance and operation of new EEMs.

In practice unless there is a visible and transparent return on investment or legislative requirement to act, most buildings are built to standard building regulations which differ by country. For example, UK buildings are required to consider energy efficiency and sustainability in line with Part L of the 2013 Building Regulations. In comparison Chinese buildings are required to conform to GB51089:2005 Design Standard for Energy Efficiency, Babtiwale and Kate (2013). As a result, the uptake of truly sustainable buildings is limited. However, modern approaches to design and in particular Modern Methods of Construction (MMC) are becoming more popular and have enabled the successful construction of more sustainable buildings at cheaper cost due to scalability.

1.2.1 Modern approaches to design

Modern methods on building construction do take consideration of sustainability and energy efficiency in design. MMC in particular focus on production efficiency, quality and sustainability in line with building regulations. This is achieved through prefabrication and offsite manufacture in controlled environments. It should be noted that building using this method does limit creativity in design.

MMC include the following types:

- volumetric – three-dimensional units produced in a factory, fully fitted out before being transported to site and stacked onto prepared foundations;
- panelised – flat panel units built in a factory and transported to site for assembly into a three-dimensional structure or to fit within an existing structure;
- hybrid – volumetric units integrated with panellised systems;
- subassemblies and components – larger components that can be incorporated into either conventionally built or MMC dwellings.

Enabling MMC for construction can and does result in enhanced quality and environmental performance. MMC systems are generally produced as much as is possible in factory environments. Therefore, much less waste is created because the environment is much more controlled. The systems of MMC are usually much easier to fix, meaning less energy is used on site. Conversely, larger loads are required to be delivered to sites and therefore the transportation impact on the environment is greater.

Another MMC that is increasing in popularity is the use of Structurally Insulated Panel System (SIPS). These boards can provide an exceptionally low U value (the measure of how effective a material is as an insulator) in the external wall, meet all the requirements of building regulations and provide as durable a building as traditional systems. MMC applies the fabric first approach to design and does enable the production of a sustainable building in a controlled environment. For large scale new modular construction this is the new modern preferred technique. One specific example of the use of SIPS is the Huf Haus, see Figure 1.5. The construction can be made available as zero-energy, requiring little heating and the use of solar photovoltaic systems to minimise the use of grid electricity.

Figure 1.5 Huf Haus

Although SIPS and MMC offer a modern sustainable solution to new building, they do not address the issue of sustainable retrofitting of existing buildings. This is a more complicated area requiring a different level of intervention and financial analysis. For existing buildings, government legislation and legal requirements have a greater role to play in enforcing change.

1.2.2 Legislative and legal requirements

During recent years there has been increased effort to support the retrofit of older building stock and production of new sustainable buildings. Governments across the world have introduced new legislation and provided funding and support for such programmes. This has been met with mixed opinion as some property owners have opposed the regulations due to the cost implications of making significant improvements to their properties without seeing any return on investment. Others have embraced the changes and welcomed the opportunity to ensure that their buildings have lower energy bills and are futureproofed. Some examples are provided below:

United States

The federal government of the United States, provided financial assistance to support existing building retrofits through the US Environmental Protection Agency, EPA (2016).

Australia

The Commercial Building Disclosure (CBD) programme requires the owners of Australia's large commercial office buildings to provide energy efficiency information to potential buyers, CBD (2016).

European Union

The European Union through the Energy Performance Building Directive EPBD (2016) have mandated the requirements that led to Part L of the Building Regulations.

The 2010 recast of the EPBD sets a target for all new buildings to be 'nearly zero-energy buildings' by 2020, including existing buildings undergoing 'major renovation'. In 2007, EU leaders agreed demanding climate and energy targets, to be met by 2020. These are commonly referred to as the '20–20–20' targets. These are:

- a reduction in EU greenhouse gas emissions of at least 20 per cent below 1990 levels;
- 20 per cent of EU energy consumption to come from renewable resources;
- a 20 per cent reduction in primary energy use compared with projected levels, to be achieved by improving energy efficiency.

United Kingdom

All UK requirements stem from the EPBD and include the requirement for certain public buildings to declare their energy performance through Display Energy Certification (DEC), and the requirements for Energy Performance Certificates (EPCs). The Climate Change Act 2008 commits the United Kingdom to reducing greenhouse gas emissions by at least 80 percent over 1990 levels by 2050, requiring legally binding carbon budgets for 5-year periods, DECC (2008).

The Coalition Government had promised that all new homes would be zero carbon from 2016 and non-domestic buildings built from 2019, with earlier dates for schools (2016) and public-sector buildings (2018). Originally, the definition of 'zero carbon' was intended to include all the emissions that a building was responsible for (both regulated and unregulated energy). In 2011 the level of ambition was scaled back to exclude non-regulated energy use such as appliances and cooking. The Government consulted on changes to Part L of the Building Regulations for 2013 and recently announced the uplift for Part L.

Whether a new building is constructed using MMC or an existing building is refurbished following a legislative regulation, the aim is clear. It is to create more sustainable residential and non-residential buildings. There are of course a number of common solutions to achieving this goal.

1.2.3 Common solutions

There are a large number of factors to consider when creating and financing a sustainable building. There are certain limitations and opportunities. Moreover, the building type, size, shape, age, use and condition all have a profound effect on what can and cannot be achieved within a specific budget.

New build

Common solutions for new build are based on the fabric first approach discussed in this chapter. All new buildings are required to meet the appropriate building regulations and consider energy efficiency. As a result, new buildings are generally well insulated, double glazed and fitted with an energy efficient condensing boiler to minimise energy use. If any low or zero carbon technologies are then considered they tend to be those highlighted in Table 1.2 i.e. LED lighting and Solar PV for the reasons discussed earlier. A survey of over 200 housing associations in the United Kingdom, NHBC Foundation (2015) confirmed this to be the case.

Retrofit buildings

Sustainably retrofitting an existing building is slightly different and variable depending upon a number of different factors such as location, type, use, classification and of course cost. The most common approach is to consider a fabric first approach and as with new built the common technologies considered are LED lighting and Solar PV. Modern retrofit strategies assess heating, lighting and ventilation and are focused on the following key areas as identified in the CIOB (2011) buildings under refurbishment and retrofit guide. This information is reproduced in Table 1.3 which shows individuals elements and the most common energy saving option or EEM available for residential and non-residential buildings.

For any building there is no one fit solution and depending upon its orientation, use, age, materiality, location and thermal mass properties, different measures will yield different result. This is supported by the findings on the Retrofit Challenge: Delivering Low Carbon Buildings report, CLCF (2011). This report and others have found that *good insulation combined with an efficient heating/cooling system and the use of LED lighting with Solar PV* tends to be the most common combination of solutions.

Now that we have an understanding of the how sustainable design works in principle and practice, the key questions remain. How do the professionals work in collaboration and how do we successfully integrate the teams? The answer to these questions is simple. Any new build or retrofit project needs the complete collaboration of all involved to reach a common goal. The team should integrate from the onset, consider the options discussed above and the common solutions. Furthermore, there needs to be a fundamental change to the silo mentality of working with each team collaborating in the development of new procedures, approaches and processes. When successful, the next key stage is to then consider

Table 1.3 Modern retrofit strategies

Element	*Energy efficiency measures*
Walls	Insulation of cavities or on external/internal surfaces
Roofs	Usually loft insulation
Doors	Either replaced or draught-proofed
Windows	Either full replacement, draught-proofed or secondary glazed
Floors	Insulation
Lighting	Passive design and active design through new controls, occupancy sensors, LED and other low-energy technologies
Boilers	Replacement with high-efficiency condensing boilers with new smart control systems
Chiller plant improvement	Upgrade of plant, pumps, piping and controls
Controls	Installation of a Building Management System (BMS), upgrade to include digital controls and greater number of sensors
Air conditioning	Upgrade and provide passive replacement in areas of building where possible
Renewable energy	Photovoltaics, solar thermal hot water, wind energy, retrieved-methane powered plant installations, wood and organic-waste power-sourced heating or power plant, replacing traditional air conditioning with air-source (ASHP) or ground-source heat pumps (GSHPs), micro-hydro power
Water conservation	Low-flow water fittings and shower heads, low-flow plumbing equipment, water-efficient irrigation, greywater systems and rainwater harvesting
Electrical peak saving	Thermal-energy storage, on-site electricity generation
Advanced metering	Smart metering, half-hour metering
Distributed generation	Combined-Heat-and-Power (CHP), Combined-Cooling-Heating-and-Power (CCHP), fuel-cell technology, micro turbines

Reproduced table from CIOB (2011)

the new options to finance the measures proposed. The team should be aware of the options available and carefully select the right financial strategy that will work with that particular building.

1.3 Financing modern sustainable buildings

There are numerous methods of financing both traditional and modern existing and new sustainable buildings. Both will commonly rely on self-financing and/or a 'leveraged' loan (see Section 2.1.2) which could be supported by, for example, a government grant or subsidy.

A new or retrofit build will commonly be funded on (a) an arrears basis or (b) through an advanced staged payments scheme. Funding on an *arrears basis* works on the principle that funds are released at various stages of the life of the property; when particular elements are complete i.e. substructure or floor. This is a very familiar and common form of modern financing and one in which most individuals are comfortable with. The second type, *advanced staged payments*, is generally more expensive as it is riskier for the lender. This is because the clients receive each payment before the stage is actually commenced. Introduce the word 'sustainability' to these conventional borrowing schemes and there is an immediate increased perceived risk. Should an organisation wish to improve their energy efficiency and create a sustainable building they will typically appoint contractors, product and service suppliers. They may also appoint a sustainability or energy manager to survey the existing building or comment on the design of a new building to identify energy saving options. New energy efficient equipment may be purchased or leased and the scope of work could change. This in itself changes the plan of works and could delay completion of particular stages of the building or indeed cause unexpected problems due to the need for additional specialist teams to install equipment. All of which can raise alarm bells with the various banks and lending authorities, potentially making the loan more expensive. This fragmented approach to creating a sustainable building is not helpful and the organisation takes a 'perceived' risk of investment as the savings are only estimated. A common outcome to this is that only some of the energy savings are realised with the full potential not achieved. If, however the organisation is confident in the approach, the lending authorities are in full support of the scheme and the range of potential options; then sustainable technology can be introduced at the outset and is integral to the refurbishment or new build plans. This results in a much greater level of energy savings and the subsequent creation of a sustainable residential or commercial building.

1.3.1 Sustainable residential buildings

For the vast majority of residential buildings, the majority of people are not willing to pay a significantly higher premium to ensure longer-term energy reduction. However, a large number of people are willing to consider shorter-term lower cost energy efficiency improvements such as LED lights. In a study in the UK by Envirohome, published in CBRE (2009), they found that 30 per cent of consumers indicated a willingness to pay over £10,000 for a fully fitted 'envirohome' in comparison to a conventional home. The difficulty in paying this higher premium for a new energy efficiency home is highlighted in this report and consists of two particular reasons. First, the general public do not generally consider lifecycle costing and are unaware of discounted cashflow calculations to calculate the present value of future energy cost savings and resultant paybacks. Second, house buyers are often constrained in their ability to purchase a house by the size of the mortgage they are able to secure, which governs the absolute amount they can spend. Thus, if a purchaser has a total budget of £280,000 to spend, what would they rather purchase? A '£260,000 house' with £20,000 of green technology, which will deliver them a subsequent 'income or saving' in the form of energy cost-savings over the coming years? Or a conventional £280,000 house which in direct comparison could be bigger, better located, or with a larger garden?

Retrofitting existing residential buildings faces the same conundrum. The majority of people do not invest in energy efficiency measures for a longer-term benefit, instead they may consider a home improvement of some type or description. If an incentive is in place then that can be the driver to encourage update. A report by the UK Green Building Council, UKGBC (2013), investigated a large number of retrofit incentives for residential properties. The report identified three of the most promising options for the wider uptake of energy efficiency measures (1) variable stamp duty land tax (2) variable council tax and (3) energy efficiency FIT.

Variable stamp duty

A system whereby house buyers receive a discount if a property is above a given energy efficiency standard, or pay a higher rate if its performance is poor.

Variable council tax

Council tax rates could be varied according to the energy efficiency of a property, with discounts for high performance properties and increased rates for those with poor energy efficiency.

Energy efficiency FIT

Just as renewable energy FITs make regular payments to households for producing clean energy, an energy efficiency FIT would reward households for installing measures which would reduce their energy demand.

1.3.2 Sustainable commercial buildings

Financing sustainable commercial buildings is different from that of residential, in that a commercial property developer has different multiple objectives. They are to maximise return on investment, attract and retain the right client, look to establish a long lease with minimal break clauses and ensure that the building is fit for purpose. It is for these reasons that sustainability can be perceived as either an incentive or burden. An incentive in that a highly efficient and sustainable commercial building will attract clients as the energy costs will be lower and staff productivity should be greater. A burden in that sustainability is perceived to be costlier and the developer does not directly benefit from reduced energy costs. When considering all the pros and cons, the fact is that a sustainable commercial building is valued more highly than conventional buildings. A recent study based on data collected by the CoStar Group, (Miller, Spivey, and Florance (2007)) shows evidence that supports this claim. Within their database of around a quarter of a million commercial properties in the United States, some have had their energy efficiency rated using the US Environmental Protection Act (USEPA) Energy Star measurement. A voluntary program that helps businesses and individuals save money through energy efficiency. In a sample of 223 buildings rated using Energy Star compared with 2,077 Non-Energy Star buildings. Analysis of the samples showed that:

- The more energy efficient 'green' buildings attracted rents per sq ft that were around 6 per cent higher than traditional buildings.
- Over the 15 months analysed, the average rent on the green buildings rose by 8.2 per cent, compared with 7.6 per cent growth on the traditional buildings.
- The green buildings appeared to secure a sale price premium of around 9 per cent in 2005 and as much as 30 per cent in 2006.

Further analysis by Eichholtz, Kok, and Quigley (2008) also compared the rental difference between a sample of Energy Star and LEED-rated office buildings in the United States with nonrated buildings in the immediate vicinity. The comparison was based on actual contractual lease rents as opposed to anecdotal or engineering-based estimates. The study found that rents for green offices were about 2 per cent higher than those for comparable buildings located nearby. Effective rents, adjusted for respective occupancy levels, show a rent differential of around 6 per cent. The University of Texas at Dallas student services building was one of the first academic buildings in Texas to receive LEED Platinum states and as a result they have experienced cost and energy savings to the sum of a $60,000 annual electricity saving, see Figure 1.6.

Figure 1.6 The University of Texas at Dallas student services building

With an understanding of what sustainable buildings are and how they are constructed, in the next chapter we will consider how they can be financed.

1.4 Guiding principles

Throughout this book there are a series of guiding principles to assist the reader in making the right decisions at the right stage in the development of a sustainable building. These guiding principles are the crucial elements for success and for Chapter 1 they are listed below:

Principle 1.1 – Cost-effective sustainable buildings can and are being built but there is an additional capital expenditure.

Principle 1.2 – For all developments, the client must commit to producing a sustainable building and in doing so they commit to a higher capital expenditure for longer term gain.

Principle 1.3 – Producing a sustainable building in principle is simple. Take a 'fabric first approach' and then consider passive and active design.

Principle 1.4 – Producing a sustainable building in practice is complex. Most building owners still consider sustainable design as a 'nice to have' feature that is too capital-expenditure intensive. There are multiple and conflicting beneficiaries and a general unwillingness to invest upfront.

References

Babtiwale, E., and Kate, S. (2013). *McKinley sustainability workshops: Sustainability seen through the eyes of children.* BESS-SB13 California: Advancing Towards Net Zero. Pomona, California, USA 24th–25th June 2013, p. 130

Barlow, S., and Fiala, D. (2007). Occupant comfort in UK offices – how adaptive comfort theories might influence future low energy office refurbishment strategies. *Energy and Buildings, 39*, 837–846

BedZED. (2016). Beddington zero emission development. Retrieved from www.energy-cities.eu/IMG/pdf/Sustainable_Districts_ADEME1_BedZed.pdf (Accessed 29 February 2016)

BRE. (2016). The Building Research Establishment Green Guide to Specification. Retrieved from www.bre.co.uk/greenguide/podpage.jsp?id=2126 (Accessed 29 February 2016)

BREEAM. (2016). Building Research Establishment Environmental Assessment Method. Retrieved from www.breeam.com (Accessed 29 February 2016)

Bull, J. (2003). *Life cycle costing for construction.* Abingdon, UK: Routledge

CASBEE. (2016). Comprehensive Assessment System for Built Environment Efficiency. Retrieved from www.ibec.or.jp/CASBEE/english/ (Accessed 29 February 2016)

CBD. (2016). Commercial Building Disclosure (CBD) Program. Retrieved from www.cbd.gov.au/overview-of-the-program/what-is-cbd (Accessed 18 April 2016)

CBRE. (2009). *Who Pays for Green? The Economics of Sustainable Buildings.* CBRE EMEA Research

CIOB. (2011). Buildings under Refurbishment and Retrofit. Retrieved from www.carbonaction2050.com/sites/carbonaction2050.com/files/document-attachment/Buildings%20under%20Refurb%20and%20Retrofit.pdf (Accessed 18 April 2016)

CLCF. (2011). Retrofit challenge: delivering low carbon buildings. centre for low carbon futures. *Research Insights into Building Retrofit for the UK.* Report No. 4

CPR. (2013). Construction Products Regulation. Retrieved from www.constructionproducts.org.uk/publications/industry-affairs/display/view/construction-products-regulation/ (Accessed 29 February 2016)

DECC. (2008). Climate Change Act. Retrieved from www.decc.gov.uk/en/content/cms/legislation/cc_act_08/cc_act_08.aspx (Accessed 29 February 2016)

Dezeen. (2009). Solar decathlon house by Technische Universtät Darmstadt (TUD). Retrieved from www.dezeen.com/2009/10/16/solar-decathlon-house-by-technische-universitat-darmstadt/ (Accessed 1 September 2016)

Edwards, B.W., and Naboni, E. (2013). *Green buildings pay: Design, productivity and ecology.* Abingdon, UK: Routledge

Eichholtz, P., Kok, N., and Quigley, J. (2008). Doing Well by Doing Good? Green Office Buildings, Working Paper No W08–001, Fisher Center for Real Estate and Urban Economics, University of California, Berkeley

EPA. (2016). Budget Request Increases Support for Communities to Deliver Core Environmental and Health Protection. Retrieved from https://yosemite.epa.gov/opa/admpress.nsf/0/A4864D01268A849785257F54006E08F5 (Accessed 29 February 2016)

EPBD. (2016). Energy Performance of Buildings Directive. Retrieved from www.epbd-ca.eu/ (Accessed 9 November 2016)

EPD. (2016). Environmental Product Declaration Retrieved from www.environdec.com/en/What-is-an-EPD/ (Accessed 29 February 2016)

EuroPHit. (2016). Retrieved from http://europhit.eu/ (Accessed 25 November 2016)

GBCA. (2016). Australian Green Star Programme. Retrieved from www.gbca.org.au/green-star/ (Accessed 29 February 2016)

Green Globes. (2016). Canadian Globe Series. Retrieved from www.globeseries.com/ (Accessed 9 March 2016)

Häkkinen, T., and Belloni, K. (2011). Barriers and drivers for sustainable building. *Building Research & Information, 39*(3), 239–255, DOI: 10.1080/09613218.2011.561948

HM Government. (2010). HM Government Low Carbon Construction Innovation & Growth Team Final Report, Department for Business, Innovation and Skills, Crown Copyright 2010

Hopfe, C., and McLeod, R. (2015). *The passivhaus designer's manual. A technical guide to low and zero energy buildings.* Abingdon, UK: Routledge

Housladen, G. (2012). *Building to suit the climate.* Basel, Switzerland: Birkhauser

Hydes, K., and Creech, L. (2000). Reducing mechanical equipment cost: the economics of green design. *Building Research & Information, 28*(5/6), 403–407

Larsson, N., and Clark, J. (2000). Incremental costs within the design process for energy efficient buildings. *Building Research & Information, 28*(5/6), 413–418

Lechner, N. (2014). *Heating, cooling, lighting.* New Jersey: John Wiley & Sons

LEED. (2016). Leadership in Energy and Environmental Design. Retrieved from http://leed.usgbc.org/leed.html (Accessed 29 February 2016)

Miller, N., Spivey, J., and Florance, A. (2007). Does Green Pay Off? (Burnham-Moores Center for Real Estate, San Diego University/CoStar)

NHBC Foundation. (2015). Sustainable technologies: the experience of housing associations. Retrieved from www.nhbcfoundation.org/Publications/Primary-Research/Sustainable-technologies-NF6 (Accessed 2 March 2016)

Nelms, C., Russel, A.D., and Lence, B.J. (2005). Assessing the performance of sustainable technologies for building projects. *Canadian Journal for Civil Engineering, 32*, 114–128

Passivhaus. (2016). The Passivhaus Standard. Retrieved from www.passivhaus.org.uk (Accessed 2 March 2016)

Pinkse, J., and Dommisse, M. (2009). Overcoming barriers to sustainability: an explanation of residential builders' reluctance to adopt clean technologies. *Business Strategy and the Environment, 18*, 515–527

RICS. (2015). COP21: Built Environment crucial to attaining emissions targets. Retrieved from www.rics.org/uk/news/news-insight/press-releases/cop21-built-environment-crucial-to-attaining-emissions-targets/ (Accessed 18 April 2016)

Roberts, S. (2008). Altering existing buildings in the UK. *Energy Policy, 36*, 4482–4486

UKGBC. (2013). Retrofit Incentives. UK Green Building Council Task Group Report July 2013

UN Documents. (1987). Our Common Future. Retrieved from www.un-documents.net/our-common-future (Accessed 5 February 2017)

UNEP. (2007). *Building and climate change: status, challenges and opportunities.* Nairobi: UNEP

World Green Building council. (2013). The business case for green buildings. Retrieved from www.worldgbc.org/files/1513/6608/0674/Business_Case_For_Green_Building_Report_WEB_2013-04-11.pdf (Accessed 18 April 2016)

2030 Districts. (2016). Retrieved from www.2030districts.org/seattle (Accessed 18 April 2016)

2 Methods of financing sustainable buildings

There are various and numerous methods of financing sustainable buildings and it should be noted that prior to consideration of the options; it is essential that the client commit to creating one. Chapter 1 has discussed this prerequisite in detail explaining the roles of the individuals involved and how to consider sustainable design. This starts with the fabric first approach and the subsequent use of both passive and active design principles. We have also learnt how to reflect on the financial implications of doing so and what the guiding principles are. In this second chapter, an exploration of the traditional and new methods of financing sustainable buildings are discussed and considered.

2.1 Traditional methods of finance

The most traditional method of financing a sustainable building includes a combination of the following:

- *self-financing* – whereby a measure is paid for directly from an organisations or individuals cash reserves;
- *loans* – to provide full or part investment to bridge the funding gap i.e. conventional bank loan;
- *grants* – such as those provided by the UK government for energy efficiency programmes.

Generally speaking, the pros and cons of these traditional methods are seen below:

The pros are, NREL (2012):

- The building owner controls every aspect of the project and has autonomy over decision making.
- There is generally a direct link to energy savings.
- Paybacks are known and savings can be predicted.

The cons are:

- The client takes the full risk.
- There is a direct link to energy losses if the solutions do not work as expected.
- Capital expenditure is required and funds must be found upfront.

These traditional methods have mixed results as there are numerous financial arrangements depending upon a large number of variables.

Self-financing

For example, a building is poorly insulated and the owner wishes to make the building more sustainable by reducing heat loss through the fabric and thereby using less energy. A solution to the problem is to take the fabric first approach and properly insulate the building. There are proven techniques using for example mineral wool or foamed insulant on the interior or exterior of the property. The costs are predictable; the energy savings are accurate and the risks are very low. A *self-financing* model, in this particular case, will work as paybacks are known and evidenced through numerous successful examples.

Loans

A second example is that of a school or hospital. Historically a development of this type would have been publicly funded by the tax payer. The UK government changed that condition by introducing Private Finance Initiative (PFI) whereby the projects are put out to tender with bids invited from building firms and developers who put in the investment to build and then lease them back. Lease arrangements for PFI projects are long term, often 25 years or longer. This is a *loan* arrangement with typically a high rate of interest.

Government grants

A final example is that of a government grant. In this example a building owner wishes to create a sustainable building by using heat from the ground. A new Ground Source Heat Pump (GSHP) can be used to supplement heating to the property thereby reducing the amount of gas used in boilers, saving the owner money and reducing the overall carbon footprint. GSHPs vary by design, will offer different energy savings dependent upon the site, and will vary in cost according to size and ground conditions. In this example, there are a lot more unknowns and as a result a traditional funding method may not be the best solution. There is more risk involved and the return on investment is less accurate. In addition, there is a new element of complexity and it becomes increasingly difficult for the building owner to make a decision. The owner may therefore choose to *self-finance* the scheme but seek support from a government-backed *Grant*. In this case the

Government can provide some level of protection against any unforeseen problems as they will only use approved suppliers and contractors with the necessary professional indemnity and insurance guarantees. No two projects are the same and any number of financial arrangements can be made using a combination of the above methods. Further details of which are provided below.

2.1.1 Self-financing

Self-financing, in this context, refers to a method of funding the construction of a new building or refurbishment of an existing building without the need for additional borrowing or grant support. Very few people are in this unique position and for those individuals that are; they do so because they are familiar with the processes, aware of the risks involved and have contingency plans in case for any unforeseen circumstances. Self-financing a sustainable building however is different. There are a lot more unknowns and the level of risk involved increases, especially if the building owner is not a technical specialist or does not have a team of specialists to advise correctly. In addition, there are many different types of sustainable building, some of which have minimal features while others consider every facet and become certified to particular standards i.e. Passivhaus. There are also very few examples of 100 per cent self-financing sustainable buildings and the vast majority will use some form of additional borrowing through some combination of loan and/or grant funding.

2.1.2 Loans

Most building owners and/or property developers (who are unable to completely self-finance a project) will borrow or 'leverage' money to enable large-scale construction and/or refurbishment to take place. For a sustainable building this method of borrowing is even more prevalent, as most building owners do not have the additional funds available for longer-term paybacks. A common thread from both a conventional and sustainable building is that the majority of owners will seek a return on investment. A mortgage is a good example and most individuals will attempt to '*remortgage*' on occasions to seek a better financial deal and resultant return on investment. The same concept applies for a sustainable building and in particular the use of so called energy efficiency measures (EEMs) as a method of realising that saving. An example of leveraging is given below with excerpts from Investopedia (2016).

Leveraging

In this example, a property is purchased at £500,000 and a deposit of 20 per cent (£100,000) is made. The buyer is essentially using a relatively small percentage of his or her own money (self-financing) to make the purchase, and the majority of that capital outlay is being provided by the lender.

Assuming the property appreciates at 5 per cent per year, the borrower's net worth from this purchase would grow to £525,000 in just 12 months. If the borrower decided to purchase a property at £100,000 and assuming the same 5 per cent rate of appreciation, the buyer's net worth from this purchase would have increased by £5,000 over the course of 12 months versus £25,000 for the more expensive property. The £20,000 difference demonstrates the potential net worth increase provided through the use of 'leverage'. Therefore over the long term the use of leverage can have a significant, positive impact on net worth. It must be noted that there are of course interest payments to make on the mortgage.

Conversely leveraging can work against the borrower in that if the property depreciates by 5 per cent per year for several years in a row, the leverage works in reverse. After year one, the £500,000 property could be worth £475,000. In year two, it could be worth £451,250 – a loss in equity of £48,750. Under that same 5 per cent price-decline scenario, if that £100,000 had been used for an all-cash purchase of a £100,000 home, the buyer would have lost just £5,000 the first-year home prices fell. Almost always leveraging involves the risk that borrowing costs will be larger than the income from the asset, or that the value of the asset will fall, leading to incurred losses.

If one now considers a sustainable building and the use of EEMs the same rules apply and we simply replace the value of the asset by the amount of energy used or saved. For example, a client spends £500,000 per year on energy in their building and they do not have enough cash reserves to fund a large-scale energy reduction programme. If they borrow the necessary monies through a loan they can then implement the programme and could achieve significant savings. If for example they installed a range of EEMs that could save them 5 per cent per year (£25,000) and the capital and operational expenditure of the measures was £50,000. Then after 24 months the return on investment is made and each year thereon the client is saving a further £25,000 per year (with all other variables remaining static of course). Conversely the EEMs may not perform as expected if they yield no energy savings then the energy bill rises to £550,000 in year one. As with the example above there would be interest payments to make on the energy reduction loan, therefore this figure is not entirely accurate but is useful as an indicator of the potential benefits that a loan can provide.

Given that borrowing through a loan can be a risky venture if the client is not a technical expert in selecting and using EEMs, the next option is to consider the use of the various government grants and schemes.

2.1.3 Government grants and schemes

There are a large number of government-led schemes linked to the construction and/or retrofit of a sustainable building. Generally speaking, these schemes have not been as successful as originally envisioned due a variety of different reasons. However, they are still used by a large number of organisations and individuals. Specific examples of common government-led energy-efficiency schemes include the Malaysian Green Technology Financing Scheme, GTFS (2016), the European Investment Bank, EIB (2016), the European Bank for Reconstruction and Development, EBRD (2016), the Green Investment Bank GIB (2016) and the UK Green Deal (2016). The details of which are provided below.

Green Technology Financing Scheme

In 2010, the government of Malaysia announced its Green Technology Financing Scheme (GTFS), a loan incentive to attract innovators and users of green technology. As a result of this incentive, there has been an increasing trend of companies building green, Diyana and Abidin (2013), with the rate of green building certification in Malaysia rising from just 1 to 137 between 2009 and 2013, Aliagha, Hashim, Olalekan Sanni, and Ali (2013).

European Investment Bank

The European Investment Bank (EIB) offers long-term finance in support of investment projects. It advises and assists in the application of European Union (EU) funds through the JASPER (2016), JESSICA (2016) and ELENA (2016) initiatives. The EIB can also offer finances on individual projects through loans, technical assistance, guarantees and venture capital (finance that the EIB can provide for startup companies that it feels have long-term growth potential).

European Bank for Reconstruction and Development

The European Bank for Reconstruction and Development (EBRD) operates from Central Europe to Central Asia and invests in global projects that would otherwise be unable to attract funding. This includes funding for cities to implement sustainable energy and energy efficiency improvements in buildings, SCI (2011).

Green Investment Bank

For large organisations such as local authorities or NHS trusts, the Green Investment Bank, GIB (2016) provides so called 'Green Finance' to fund for example building retrofits (e.g. lighting, insulation, glazing) and/or onsite energy generation (e.g. Combined Heat and Power [CHP], renewable heat, heat pumps). In short, the GIB offers discounted interest rates over a long time period to promote the uptake of energy efficient technologies. The aim of which is to promote UK wide carbon and energy reduction. A report by Vivid Economics (2011) for the Department for Business, Innovation & Skills made the case for the GIB as an enduring but responsive institution, offering real value for money. Others are less supportive as the government plans to privatise the GIB in the future, which may influence the competitive low interest rates currently on offer for renewable schemes.

Carbon and Energy Fund

The Carbon and Energy Fund (CEF) was launched in 2011 and was specifically created to fund, facilitate and project manage complex energy infrastructure upgrades for the NHS and wider public sector. The following information is taken from the CEF (2016) website. The CEF itself helps public sector organisations by:

- providing advisors to help the Public Sector with their projects for the entire life of the project;
- giving member organisations the use of the CEF's proven EPC, that transfers risk and guarantees savings;
- giving member organisations usage of the CEF framework procurement model, so that procurement is quick, with a significantly reduced chance of challenge;
- providing a source of funding if the member organisation does not want to use its own funding, or when the bidder cannot provide better funding or grants to support the project;
- supporting the member organisation's team by overseeing the installation phase of the project and certifying practical completion;
- auditing the project monthly, quarterly and annually so as to prove the delivery at all times during the contract life;
- providing legal support for the process of making the standard project agreement, project specific to the member organisation's needs;
- recovering all CEF costs from within the contractor payment, so that the member organisation faces no advisor bills, and only has to pay for any part of the project when the project is installed, proven and handed over to operations; and
- managing the CEF framework leading to fewer procurement issues and contractor problems.

Green Deal

This was the UK government's flagship energy saving program and in principle was simple. In short, the energy savings produced from installing a new EEM would be used to cover the installation and project cost, Green Deal (2014).

Have these types of schemes been a resounding success? There has been mixed success and in this example we can investigate the UK Green Deal. As mentioned above this was the UK governments energy saving program whereby the energy savings would be used to cover the costs. In reality however, if one obtained a Green Deal loan, it was just that (a loan) with an interest rate that was comparatively high (6.96 per cent). The application process was lengthy and customers could only use an approved Green Deal installation company. As a result, the entire scheme was (a) more expensive than it should have been (b) difficult to administer and (c) confusing for the customer. In July 2015, the Office of Gas and Electricity Markets (OFGEM) decide to close the scheme. Examples like this receive unwanted media attention and add to the common notion that sustainable buildings are too costly and risky. However not all government schemes and incentives are akin to the Green Deal and some of the schemes identified above are great success stories that have been implemented across the world. Additional good examples include the:

- US Renewable Portfolio Standards RPS (2016) designed to increase the generation of electricity from renewable resources;
- Chinese Corporate Income Tax (CIT) reduced rate for advanced and new renewable energy technology enterprises;
- Indian Generation Based Incentives (GBI) scheme to attract foreign investors into renewable energy;
- Polish Green Certification System (GCS) with remunerates for renewable energy production;
- UK Renewable Heat Incentive (RHI), which for renewable heat technologies, households or businesses installing small-scale renewable heating systems will be paid a fixed amount based on the amount of heat the system was estimated to produce; and
- globally used feed in tariffs (FITs) which encourage home-owners and businesses to install renewable technology by guaranteeing a long-term premium payment for electricity generated and fed into the grid.

In addition to the above there are addition private-sector methods of obtaining third-party support through for example Siemens Financial Services (SFS).

Energy Efficiency Financing

Energy Efficiency Financing (EEF) is a partnership set up by the Carbon Trust and SFS. The scheme is designed to help facilitate investment in new technology for buildings in the forms of leases, loans and hire purchase. The cost of the equipment and installation is designed to be offset by energy cost savings. The finance solutions available through the EEF scheme are usually based on one or a combination of a finance lease, lease purchase, loan finance or, in some cases, an operating lease.

Finance lease

With a finance lease, title to the equipment remains with SFS throughout. The customer makes a series of 'rental' payments during the minimum term of the agreement, which will include interest and charges, and the rental payments can be flexed to suit your cash flow. When the minimum term ends, the customer chooses to continue hiring the equipment or return it to SFS. A finance lease enables the customer to upgrade equipment during the term and can also include maintenance, so that both the equipment and its support services are paid for under a single contract.

Lease purchase

Lease purchase is a highly flexible way of financing equipment. It enables you to spread the cost of purchasing the asset across the agreement period. You may then choose whether to purchase the item outright at the end of the agreement. In contrast to finance lease, Value Added Tax (VAT) on lease purchase is paid in one lump sum before the finance agreement begins.

2.2 New methods of finance

As discussed above the traditional methods of financing conventional and sustainable buildings have received mixed results and as a general rule of thumb the traditional methods work well for small individual buildings or small-scale development sites. For larger schemes and masterplans an alternative financial structure is more effective. It is the author's opinion that there is a need to change the status quo and embrace change in order for the mass uptake of sustainable buildings to become a reality. Some of the market leaders in this area have already implemented change and are currently using the new methods discussed in this chapter.

This new financial structure, which is growing in interest, popularity and frequency, is that of EPC and in particular the use of ESCOs. In short, the EPC model benefits energy end-users by reducing the technical risk they would otherwise incur through a traditional method of finance with the ESCO providing the engineering based solution. The full details are which are provided below.

2.2.1 Energy performance contracting

As discussed in an article written in 2degrees (2013) and reworded here, an EPC is a partnership between a customer and an ESCO that allows the improvement of building energy efficiency without any upfront capital costs to the end client. Typically with an EPC, the ESCO will introduce a number of EEMs. But what is different about an EPC compared to a normal programme of building upgrades is that the provider will guarantee that the energy savings delivered will pay for the capital investments in new equipment. The idea is that energy cost savings will exceed the cost of repaying the capital. The modern EPC contract is adapted from the original contracts used in the US public sector in the 1970s and 1980s which were widely promoted around the world by the US government and taken up enthusiastically in nearly all countries by entrepreneurs, policy makers and international finance institutions keen to improve energy efficiency. However, they have not been as successful as originally envisioned and in many countries, the EU and international financial institutions continue to actively promote it. In the United Kingdom, the Energy Managers Association, the Greater London Authority and the GIB advocate the use of EPCs in various ways. In short, the EPC contractor (ESCO) identifies appropriate EEMs, develops them, builds them, provides a guarantee that a set level of energy savings will be achieved, and in some cases arranges financing to pay for the projects with repayments less than the savings.

An important barrier to uptake is that EPCs are inherently long-term and complex agreements. They need to be long-term, typically between 7 and 15 years to ensure savings exceed payments. They also need to take account of all the changes that can occur over the long-term including variations in energy use caused by known factors and how wholesale changes in the way that buildings are used. Methods of Measurement and Verification (M&V), preferably by independent M&V specialists, need to be agreed and put in place. Contracts tend to be long, complex and hard to negotiate and transaction costs, especially legal costs, can be high. Clients need to dedicate significant amounts of focused resources on developing and concluding the deal, as well as ongoing management of the contract once it is in place.

Although there is a principle of *shared savings*, people considering using EPCs need to consider the true incentives of the contractor. The contractor develops and implements the project and gets paid for its work on completion and of course on that work it makes a profit margin. Fundamentally the contractor is incentivised to maximise capital expenditure and margins. That margin is not typically visible and because of the nature of the typical EPC, which will include several energy efficiency projects, it can be nearly impossible to determine the real project costs unless transparency is made part of the process. When ESCOs have been used in the commercial office market this lack of transparency, along with issues such as the split incentive (discussed in Chapter 4) between landlords and tenants, have severely hampered progress.

The typical EPC produces a long stream of savings over the contract life and after repayments of capital there is a relatively small net saving to the client. So, given that EPCs are complex, hard to negotiate and procure and produce a small net saving it is perhaps not surprising that EPC growth has been slower than hoped for in all countries. Although that option is changing, EPCs are now more commonplace. They are most suited to the public sector and buildings with an owner-occupier rather than the commercial rented building sector and they can bring much needed investment into buildings where major upgrades of plant and even building envelope are needed. They are effectively a type of Public Private Partnership (PPP) which brings new and upgraded infrastructure and reduces operating costs. Organisations looking to procure EPCs need to fully understand the contract and the allocation of risks, the incentives on both parties as well as the potential pitfalls, and appreciate the resources needed to develop and negotiate the deal before they start on the journey.

The key steps in an EPC will typically include:

Step 1: Identifying resources and funding approach – seeking approvals, establishing a team and enabling funding are all necessary starting points.

Step 2: Tendering phase – a plan for procurement is required together with a process for competitive tendering.

Step 3: Full Investment Grade Proposals (IGPs) – this is typically produced by the ESCO and should detail the range and type of EEMs to be installed.

Step 4: Install energy efficient measures (EEMs) – when the IGP is agreed the ESCO can be appointed to proceed with the installation.

Step 5: Service delivery and performance monitoring – once installed an appropriate Monitoring, Reporting and Verification (MRV) plan is required. If underperformance is identified, then remedial action can take place.

Further details and explanations can be found in the Guide to Energy Performance Contracting Best Practice, DECC (2015).

2.2.2 Energy service companies

An ESCO, in its simplest form, is defined as an 'entity providing a broad range of comprehensive energy solutions'. The ESCO model provides the finance and technical expertise to achieve energy savings. They provide an energy service rather than simply supplying a specific EEM, which has worked well in the United States for the past 20 years, Fawkes (2013). They are generally classified into the following four categories, Langloise and Hansen (2012) and a number of working case studies are presented at the end of this book.

1. *Independent ESCOs* – these are ESCOs not owned by an electric or gas company, an equipment manufacturers or an energy supply company.
2. *Building equipment manufacturers* – these are ESCOs owned by building equipment or controls manufacturers. Many will have parent companies and act as private companies.
3. *Utility companies* – these are ESCOs owned by regulated or non-regulated electric or gas companies.
4. *Energy or engineering companies* – these are ESCOs owned by international oil companies, non-regulated energy suppliers or large engineering firms.

The choice of ESCO is important and each offer will differ, therefore it is important to understand how they are structured and how this may or may not impact on the EEMs suggested. Independent of the choice all ESCO models will typically start with an energy audit to detail the buildings current energy performance and investigate how energy is used. The next step is to identify numerous EEMs and options for the procurement of energy reduction services. From the initial analysis, the ESCO will then provide a feasibility study which aims to cover the potential measures proposed and the level of guaranteed savings expected from the installation. The ESCO can provide this guarantee as they have confidence in their products, Baechler and Webster (2011). Only measures that are proven to work are suggested by the ESCO and as a result they take onboard the risk. In return they share in the savings made, through either guaranteed, shared saving or no guaranteed saving models, British Gas (2012).

Shared saving model

These types of models are less frequent and the approach involves the cost savings from the project to be shared between the building owner and the ESCO. The schedule arrangement of payments to the ESCO will vary but may be a:

- fixed percentage of savings;
- minimum fee;
- share of the savings made by the efficiency measure.

It could also be arranged so that a scaled fee is established which decreases over a certain period, as the ESCO receives its return on investment. The ESCO will typically provide the capital investment and assumes a significant percentage of the risk. Unlike a guaranteed savings model both the ESCO and the owner may experience additional monetary benefit if the savings made exceed the predicted estimates. In addition, the ESCO and building owner will share the risk, see Figure 2.1.

Guaranteed saving model

These are generally the most commonly used form of performance contracts and are characterised by the following:

- The amount of energy saved is guaranteed (with parameters).
- A fixed term with a fixed payment schedule upon which the particular ESCO ensures the savings will meet or exceed a minimum level. If the savings are not met the building owner will still receive the fixed payment.
- The contract is typically financed by the ESCO but may require capital investment from the building owner.

In a guaranteed saving model, the client secures a loan from a financial institution and pays the ESCO for services according to the terms of the EPC. The savings from the ESCO are guaranteed and as such the client can make regular and predictable payments back to the financial institution that loans the money in the first instance, see Figure 2.2.

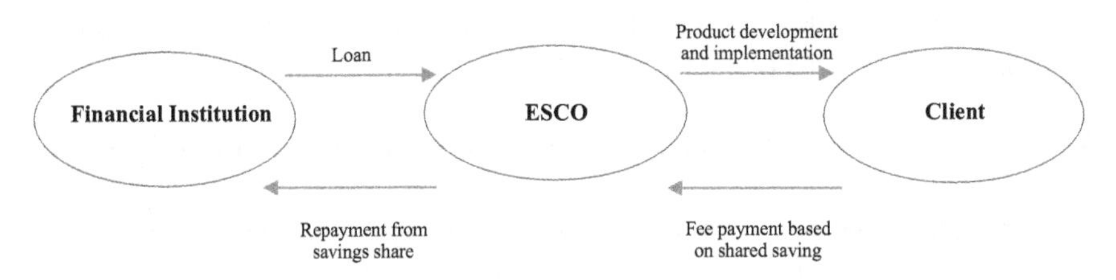

Figure 2.1 Shared saving model

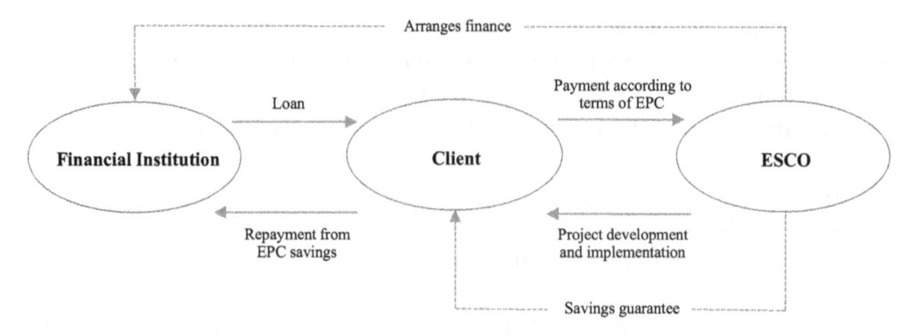

Figure 2.2 Guaranteed saving model

No guaranteed saving

In this type of contract, the ESCO does not provide the building owner with a guarantee upon energy savings or performance of the project. The contract is characterised by:

- The building owner bears the burden of the project risks.
- The building owner will provide the necessary capital either directly from cash reserves or self-financing.

In all three models the ESCO provides the initial energy audit, the EEM, installation, maintenance, and disposal/reuse. Key to the success of an EEM is ongoing measurement, reporting and verification combined with quality assurance features, Numark Associates (2011) to ensure that the proposed measures perform as initially predicted. A key feature and concern is the length of contract and time it will take to repay the investment. This can be unpredictable which is why a variable contract term is common.

Variable contract term

The ESCO designs, finances and implements the project and verifies the energy savings. If the savings are less than expected, the contract term can be extended to allow the ESCO to recover its full payment. In the so called 'first out' variation of this, the ESCO takes all the savings until it has received its full payment.

The pros of an ESCO model can be summarised as follows, Financial Institutions Group (2014) and SEAI (2016):

- Capital expenditure in some models is not required and in others can be negotiated.
- Immediate upgrade of your facilities to reduce operating costs.
- Risks are minimal as they are covered by the ESCO.
- Maintenance, repairs and installation are provided.

The cons of an ESCO model can be summarised as follows, Marino, Bertoldi, Rezessy, and Boza-Kiss (2010), ManagEnergy (2013) and Bertoldi and Rezessy (2005):

- Energy savings are shared with the ESCO.
- Some do require some initial Capital Expenditure.
- There are numerous ESCO providers on the market.
- There can be a long-term commitment with the ESCO.
- Each ESCO has preferred suppliers and particular EEMs.

Private sector vs public sector

There are differences in the type of ESCO applicable to a client in the private or public sector. A private sector client will test the market, seek numerous quotes from various ESCO providers and make a decision primarily focused on a financial return. There are numerous options available to a private sector client but no government-backed interest-free option. This is different for a public sector client who may, if they wish, employ a private sector ESCO or instead consider a government-backed provider. One such provider is Salix Finance Ltd (2016). They are proving to be the most popular choice as they are able to provide government-backed 100 per cent interest-free capital. In short, the money borrowed is used to reduce the clients' energy costs by enabling the installation of modern, more energy efficient technologies. Salix recognise that upfront capital is a common barrier for public sector organisations seeking solutions that cut their energy consumption. As Salix is a not-for-profit organisation funded by the Department for Energy and Climate Change (DECC), the Department for Education (DfE), the Welsh Assembly Government (WAG), the Scottish Government and Higher Education Funding Council for England (HEFCE), there are numerous and obvious financial benefits.

To date, Salix has funded over 14,400 projects with 1,460 public sector bodies, valued at £462.9 million. This has saved the public sector over £116 million annually and £1.7 billion over the projects' lifetime. In the future, this will reduce public sector carbon dioxide emissions by 613,793 tonnes annually and over 8.6 million tonnes over the lifetime of the projects.

There are two types of Salix funding programmes available. The first is the Salix Energy Efficient Loans Scheme (SEELS) – as an example: a school borrows £10,000 to put in new lighting and a new boiler which will save the school £2,000 per annum from reduced gas and electricity usage. For the first 5 years these savings are used to payback the interest-free loan. Once the loan is repaid, the continued savings enable the school to use the capital for other budgets, such as the purchase of equipment.

The second programme available is the '*recycling fund*', which is a ring-fenced fund managed by the public sector organisation, with money provided by the organisation and match funded by Salix. The project loan is repaid into the fund from the financial savings delivered by the projects – this allows the fund to be continually used for energy efficiency projects, hence the term '*recycling fund*'. At the same time the organisation continues to benefit from the savings that accumulate once the project has been fully repaid.

Over 120 technology types are supported by the funding programmes, some of which include building energy management systems, cavity wall insulation, CHP systems, evaporative cooling, heat recovery systems, LED lighting, lighting controls, loft insulation, pipework insulation, T5 lighting and variable speed drives.

Salix funding includes all public sector organisations and across their whole estates, including schools, higher and further educational institutions, emergency services, hospitals, leisure centres, local authorities, the NHS. Specific case studies are provided at the end of this book.

The key question remains in that how does a building owner decide upon a traditional self-financing model or new method of financing using ESCOs? The largest overriding factor dictating the decision to adopt a particular EEM as a technical, detailed and large-scale energy solution is their *perception of risk and uncertainty*. Those with the available capital and a team of experts will self-finance the project and are comfortable with the level of risk and uncertainty. Those without should opt for the use of an ESCO. For example, Salix funding is a popular type of ESCO in the public sector. This applies to the majority of building owners who across all market sectors, will predominately measure risk on the basis of finance and the level of uncertainty for the wholesale integration of EEMs. In taking a holistic approach to energy efficiency, key to the use of ESCOs is that the burden of the contractual risk rests solely on the investor as opposed to the client. Given the emphasis is on the ESCO to provide an effective and efficient energy service, clients and indeed building owners will require a satisfactory level of confidence that such acquisitions posed by the ESCO have been made in their best interest and that the procurement route adopted for the facilitation of EEM's within their building have a proven 'track record'. Having concluded upon a particular EEM that is viable, the building owner, having made an informed interdisciplinary decision based on findings from the use of for example an Energy Efficiency Value Matrix (EEVM), developed by Finnegan, dos-Santos, Chow, Yan, and Moncaster (2015). Given that predominantly building owners are not experts in such fields of energy efficiency finance, the majority may be unaware of the array of ESCOs operating within the energy sector. Table 2.1 below, which is *not* an exhaustive list, details a list of potential ESCOs currently offering EPC's to clients in both the residential and commercial markets.

Table 2.1 List of Energy Services Companies (ESCOs)

ANESCO	Balfour Beatty Workplace
Bouygues Energies & Services FM (UK) Ltd.	British Gas
COFELY Ltd.	Digital energy
EDF Energy Customers Plc.	E.ON Energy Solutions Ltd.
EEVS Insight Limited	EuroSite Power Limited
Honeywell Control Systems Ltd.	Imtech Technical Services Ltd.
Inenco	IVEES
MCW	MITIE TFM Ltd.
Norland Managed Services Ltd.	Power Efficiency
Schneider Electric	Siemens Automation & Drives
Skanska Construction UK Ltd.	t-mac Technologies
The Energy Brokers	BELIMO Automation AG
BROEN A/S	Comap SA
Danfoss A/S	Delta Dore SA
Distech Controls	Frese
HAGER CONTROLS SAS	TA Heimeier GmbH
Johnson Controls, Inc.	Kieback&Peter GmbH & Co, KG
LOYTEC electronics GmbH	Oventrop GmbH & Co. KG
Priva B.V.	Saia-Burgess Controls AG
Fr. Sauter AG	Schneider Electric Buildings AB
Theben AG	Trend Control Systems Ltd.
Tridium Europe Ltd.	

2.3 Guiding principles

The guiding principles relevant to Chapter 2 are listed below:

Principle 2.1 – Self-financing EEMs is the best option if the monies are available and a team of internal experts exist who are not concerned by the potential changes in procedure and risk. Government schemes can be useful to assist in financing EEMs however they should not be a major contributing factor.

Principle 2.2 – An important principle of an EPC is that energy efficiency investments are paid for over time by the value of energy savings achieved.

Principle 2.3 – The use of an ESCO is key in achieving a cost neutral zero carbon development for those who do not have the available capital and are concerned over the perceived risk.

References

Aliagha, G., Hashim, M., Olalekan Sanni, A., and Ali, K. (2013). Review of green building demand factors for Malaysia. *Journal of Energy Technology Policy*, *3*(11), 471–478

Baechler, M., and Webster, L. (2011). "A Guide to Performance Contracting with ESCOs" U.S. Department of Energy, Energy Efficiency and Renewable Energy, Building Technologies Program. Retrieved from www.pnnl.gov/main/publications/external/technical_reports/PNNL-20939.pdf (Accessed 7 March 2016)

Bertoldi, P., and Rezessy, S. (2005). "Energy services companies in Europe," European Commission, Director-General, Joint Research Centre, Institute for Environment and Sustainability, Renewable Energies Unit, EUR 21646 EN. Retrieved from www.energetska-efikasnost.ba/Publikacije/Literatura/ENERGY_SERVICE_COMPANIES_IN_EUROPE.pdf (Accessed 7 March 2016)

British Gas. (2012), Using Energy Performance Contracts (EPCs) to unlock energy savings and improve the patient/working environment. Retrieved from www.eoecph.nhs.uk/Files/Sustainability/EnergyPerformanceContractsUnlockingEnergySavings.pdf (Accessed 7 March 2016)

CEF. (2016) Carbon and Energy Fund. Retrieved from www.carbonandenergyfund.net/ (Accessed 1 September 2016)

DECC. (2015). Guide to energy performance contracting best practice. *Department of Energy and Climate Change*. URN: 15D/011

DECC. (2016) Feed in Tariff Review. Retrieved from www.ecolutionrenewables.com/blog.php?item=56#blog (Accessed 2 March 2016)

Diyana, A., and Abidin, N. (2013). Motivation and expectation of developers on green construction: a conceptual view. In *Proceedings of World Academy of Science, Engineering and Technology (No. 76, p. 247). World Academy of Science, Engineering and Technology (WASET)*

EBRD. (2016) European Bank for Reconstruction and Development. Retrieved from www.ebrd.com (Accessed 7 March 2016)

EIB. (2016) European Investment Bank www.eib.org (Accessed 7 March 2016)

ELENA. (2016) European Local Energy Assistance. Retrieved from www.eib.org/elena (Accessed 7 March 2016)

Fawkes, S. (2013), "Energy performance contracts: Too good to be true?" Retrieved from www.2degreesnetwork.com/groups/2degrees-community/resources/energy-performance-contracts-too-good-be-true/ (Accessed 7 March 2016)

Financial Institutions Group. (2014). Energy Efficiency- the first fuel for the EU Economy, How to drive new finance for energy efficiency investments. Part 1: Buildings (Interim Report). Retrieved from www.unepfi.org/fileadmin/publications/investment/2014_fig_how_drive_finance_for_economy.pdf (Accessed 7 March 2016)

Finnegan, S., dos-Santos, J.D., Chow, D.H.C., Yan, Q.O., and Moncaster, A. (2015). Financing energy efficiency measures in buildings – a new method of appraisal. *International Journal of Sustainable Building Technology and Urban Development*, *6*(2), 62–70

GIB. (2016). Green Deal: energy saving for your home. Retrieved from www.gov.uk/green-deal-energy-saving-measures/overview (Accessed 1 September 2016)

Green Deal. (2014). Retrieved from www.gov.uk/green-deal-energy-saving-measures/overview (Accessed 1 September 2016)

GTFS. (2016). Green Deal: energy saving for your home. Retrieved from www.gov.uk/green-deal-energy-saving-measures/overview (Accessed 1 September 2016)

Investopedia. (2016). Increasing your real estate net worth. Retrieved from www.investopedia.com/articles/mortgages-real-estate/10/increase-your-real-estate-net-worth.asp (Accessed 7 April 2016)

JASPER. (2016) Joint Assistance to Support Projects in European Regions. Retrieved from www.jaspers-europa-info.org (Accessed 7 March 2016)

JESSICA. (2016). Joint European Support for Sustainable Investment in City Areas. Retrieved from www.eib.org/products/blending/jessica/index.htm (Accessed 7 March 2016)

Langlois, P., and Hansen, S. (2012). *World ESCO Outlook.* Lilburn, GA: CRC Press

ManagEnergy. (2013). "Barriers and Drivers for Energy Performance Contracting in Europe- what is the feedback from the ManagEnergy workshops?" Stakeholder views on EPC development, Based on ManagEnergy Events in Croatia, Czech R. and Denmark. Retrieved from www.managenergy.net/lib/documents/810/original_EPC_Article_Stakeholder_feedback_from_ManagEnergy_events.pdf (Accessed 7 March 2016)

Marino, A., Bertoldi, P., Rezessy, S., and Boza-Kiss, B. (2010). "Energy services companies Market in Europe- Status Report 2010-" Institute for Energy, European Commission, Joint Research Centre, JRC Scientific and Technical Reports, EUR 24516 EN. Retrieved from http://iet.jrc.ec.europa.eu/energyefficiency/sites/energyefficiency/files/escos-market-in-europe_status-report-2010.pdf (Accessed 7 March 2016)

NREL. (2012). National Renewable Energy Laboratory "Residential Solar Photovoltaics: Comparison of Financing Benefits, Innovations, and Options". Technical Report NREL/TP-6A20–51644. Retrieved from www.nrel.gov/docs/fy13osti/51644.pdf (Accessed 7 March 2016)

Numark Associates. (2011). Measurement and Verification of Energy Savings. Retrieved from www.numarkassoc.com/measurement-and-verification-of-energy-savings.html (Accessed 7 March 2016)

Salix Finance Ltd. (2016). Retrieved from www.salixfinance.co.uk (Accessed 25 August 2016)

SCI. (2011). Financing and Contracting Sustainable Construction – innovative Approaches. Sustainable Construction and Innovation Network. Retrieved from www.sci-network.eu/fileadmin/templates/sci-network/files/Resource_Centre/Reports/Financing_and_Contracting_Preliminary_Report.pdf (Accessed 7 March 2016)

SEAI. (2016). "A guide to Energy Performance Contracts and Guarantees" Version: Draft for Consultation, Sustainable Energy Authority of Ireland. Retrieved from www.seai.ie/Your_Business/Public_Sector/Energy_Performance_Contacts_and_Guarantees.pdf (Accessed 7 March 2016)

Vivid Economics. (2011). In association with McKinsey & Co, The economics of the Green Investment Bank: costs and benefits, rationale and value for money, report prepared for The Department for Business, Innovation & Skills, October 2011

2degrees. (2013) Energy performance contracts: Too good to be true? Retrieved from www.2degreesnetwork.com/groups/2degrees-community/resources/energy-performance-contracts-too-good-be-true/ (Accessed 10 March 2016)

3 Incentives to investment

As discussed in Chapter 2, an individual sustainable building or large-scale masterplan can be financed using either a traditional self-finance model or new third-party model i.e. using an Energy Services Company (ESCO). The vast majority of building owners/developers do not wish to self-finance sustainable buildings as they don't see the immediate return on investment. The UK House of Commons Business and Enterprise Committee, UK House of Commons (2008), stated that there is a perception that sustainable buildings are costlier and investors lack the confidence to invest. Since this 2008 publication there has however been a reversal in perception but not to the extent necessary for the wholesale uptake of sustainable buildings.

Independent of the type of financial method considered there are a vast number of financial incentives to the creation of a sustainable building and this chapter focuses on these incentives. In Chapters 1 and 2 we focused on an introduction to financing sustainable buildings and the common and new methods of finance. With this newly acquired knowledge we can now consider the specific incentives and the individual and collective roles they play.

The key incentives to investment in sustainable buildings have been highlighted in a number of key publications. For example, Hakkinen and Belloni (2011) and Ala-Juusela, Huovila, Jahn, Nystedt, and Vesanen (2006) highlighted that sustainable buildings result in lower operational costs, higher productivity of occupants and long-term benefits for the national economy. This is due to reductions in emissions and the use of natural resources. Pitt, Tucker, Riley, and Longden (2009) found that a key steering mechanism for the delivery of sustainable buildings is through both fiscal incentives and regulations. A paper by Olubunmi, Xia, and Skitmore (2016) found that incentives are not only financial and can be split into external and internal. The external incentives, which are largely provided by governments, are divided into financial and nonfinancial. Internal incentives are those that allow beneficiaries to be incentivised out of choice because of the appeal of the benefits of green buildings. UNEP (2010) commented on the following incentives: (1) *Reduced Operating Expenses* – Some of the basic attributes and processes involved in green building can result in reduced operating expenses. (2) *Reduced Risks for Insurers* – Financial institutions that insure green buildings benefit from the reduced risk profiles of the buildings' owners and tenants, as well as of the building itself. (3) *Price Premiums* – Financial institutions also benefit from green buildings' price i.e. green buildings are more valuable. (4) *Alignment with Market and Regulatory Trends* – The evolution of the policy environment at all levels of government is moving strongly in the direction of requiring green buildings and energy efficiency.

There is no single publication that describes all of the incentives to investment and this chapter serves to highlight the key benefits and those that are deemed the most relevant when considering either traditional or new financial strategies. Pearce, Ahm, and Hanmi Global (2012) created a table, reproduced in Table 3.1 which highlighted three key areas covering the so called triple bottom line of sustainability 'environmental, economic and social'.

When considering a sustainable building, it is of course essential to consider the triple bottom line and in the authors opinion it is also necessary to consider building standards and regulations. The work of Pearce *et al.* (2012) and others has been expanding to also include a new incentive on standards and regulations. Therefore, there are three general specific areas of incentive which are (1) financial (2) health and well-being and (3) standards and regulations. Each is discussed below.

3.1 Financial incentives

There is a long list of financial incentives for sustainable buildings which can be external, internal, direct and indirect. For example, government-led incentives are external and can be of direct and indirect benefit. The financial incentives include the following.

Table 3.1 Incentives for sustainable buildings

Environmental benefits

- Enhance and protect biodiversity and ecosystems
- Improve air and water quality
- Reduce waste streams
- Conserve and restore natural resources

Economic benefits

- Reduce operating and maintenance costs
- Create, expand and shape markets for green product and services
- Improve occupant productivity
- Minimise occupant absenteeism
- Optimise lifecycle economic performance
- Improve the image of construction
- Reduce the civil infrastructure costs

Social benefits

- Enhance occupant comfort and health
- Heighten aesthetic qualities
- Create new and enhanced employment and business opportunities
- Minimise strain on local infrastructure
- Improve overall quality of life

Pearce *et al.* (2012)

3.1.1 Security of supply

Most modern buildings use gas and electricity to operate. From heating and cooling through to lighting and power. Depending upon the country in question the demands for each will change, for example a colder climate will require more heating whereby a warmer climate would require more cooling. The way in which gas and electricity is provided to the building is also dependent upon the location. If for example the building is in France, the majority of electricity is derived from nuclear power. If the building is in the United Kingdom, the majority of electricity is provided by a mix of nuclear, gas-fired powered stations and renewables. Gas is provided both onshore, offshore and through imports. It is this later point that is important. Each country is seeking a '*security of supply*' for its buildings and infrastructure. Why? Because they can then control the price that is paid and there is less risk of the country facing a shortage. This is where sustainable buildings have a major role to play and one such initiative to encourage the uptake of the purchase of renewable energy is a Power Purchase Agreement (PPA). In short a PPA is a contract between the generator of power and the purchaser of that power for the supply of 100 per cent renewable energy. The PPA is typically regulated to ensure that both parties do not breach any contractual arrangement. Typically, electricity will be generated from a mix of wind, solar, biomass, anaerobic digestions, landfill gas, sewage gas, energy from waste and/CHP and supplied as 100 per cent renewable to the customer. Using PPAs can therefore ensure a security of supply.

3.1.2 Market-based mechanisms

There are numerous market-based mechanisms in place to support the production of a sustainable building. These are supported by government incentives, tax benefits and operational cost savings. At present the market was confused as some of these mechanisms have been mismanaged. A clear example of which is the UK Green Deal, which in principle was simple – the energy savings produced from installing an energy efficient measure were used to cover the installation and maintenance costs. However, the application process was arduous and the amount of finance made available was insufficient. Those market-based mechanisms that are worthy of investigation and further exploration are as follows: Warm Home Discount, WHD (2016), Energy Company Obligation, ECO (2016), Feed in Tariff, FIT (2016), Climate Change Levy, CCL (2016) and the Renewables Obligation, RO (2015). These market-based mechanisms continue to have a crucial role in the incentivisation and creation of sustainable buildings.

3.1.3 Futureproofing

Not only is it important to use less energy and consider current use; but it is essential to futureproof our buildings. Projections by the UK Department for Energy and Climate Change, DECC (2015), have estimated an overall increase in the demand for energy from all sources including oil and gas, renewables and electricity. The growth rates are difficult to assess but DECC are projected to rise by 5 per cent by 2035 compared to current usage. In addition, it is useful to note that in 2014 approximately 45 per cent of the energy supplied to the United Kingdom was from imports, primarily gas and oil, DECCa (2015). This reliance on overseas markets is of major concern due to the instability of supplies from these regions and projected increases in cost.

Clearly, a sustainable building with limited need for gas and electricity (through passive design) or one that provides its own source of power through renewable technology would not necessary be concerned with the DECC statistics. This is another major incentive for the creation of sustainable buildings.

3.1.4 Reduced operational costs

The main cost saving in the production of a sustainable building is through the reduced operational costs. Investigation by Pitt *et al.* (2009) ranked the importance of eight different drivers for sustainable buildings and found that reduced operating cost remains the single biggest driver. Additionally, work by the USGBC (USGBC, 2009a) found that green buildings function more efficiently by design and produce operating cost reduction, increase of building value, occupancy ratio and rental increases. As discussed it is well established that operational costs of sustainable buildings are lower than that of conventional buildings however, the scale of incentive is sometimes difficult to quantify. How much will a sustainable building save the occupant per month/year? How much energy is saved over the lifetime of the building? These questions and many more are (1) difficult to predict and (2) may or may not be significant at the time of purchase or lease. Ala-Juusela *et al.* (2006) claim that energy efficient buildings can offer major cost savings during operation, but this is not adequately communicated to a wide audience.

3.1.5 Reduced maintenance costs

Generally speaking, maintenance costs for sustainable buildings are lower than that of a conventional building. The acting director of the US Office of Federal High-Performance Green Buildings in the US General Services Administration, GSA (2009) assessed the performance of 8,600 owned and leased buildings in his portfolio across 50 US states. He found that the high-performing sustainable buildings provided the best value for the tax payer and the public with maintenance costs decreased by 53 per cent in comparison to conventional buildings. In addition energy consumption was reduced by 45 per cent and 39 per cent less water was used. The World Green Building Council (WGBC) business case for green building report, USGBC (2013) created a useful Venn diagram to illustrate the benefits of creating a sustainable building and more importantly who benefits. This diagram, reproduced in Figure 3.1, shows that maintenance costs are reduced and this is of direct benefit to the owner of the building and the tenant. However, there is no direct benefit to the developer.

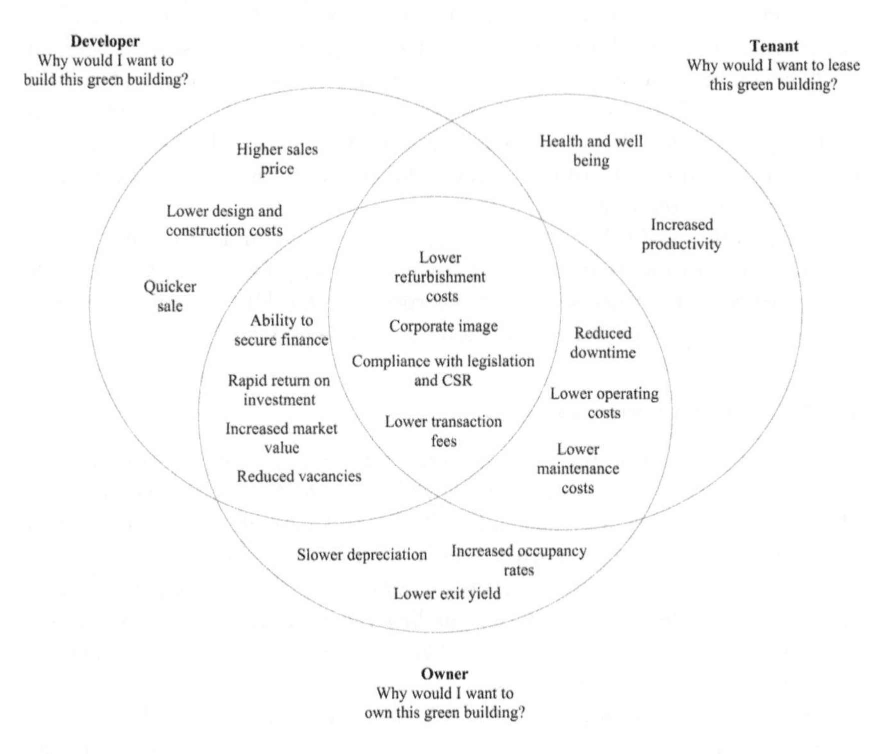

Figure 3.1 Benefits of creating a sustainable building
Reproduced from USGBC (2013)

3.1.6 Reduced payback

A US study conducted by Janzen and Brasok (2011) for the US General Services Administration (GSA) examined the cost of building a typical sustainable and conventional detached three-bedroom home in the United States. A payback of 3 years was estimated for the sustainable building. This short payback is a clear incentive to production. Payback for some can have a different meaning and in this book it refers to the time period within which capital expenditure for sustainable technology is repaid. For example, a solar thermal installation costs £10,000 and once installed that saving is recovered in 5 years as £2,000 is saved on energy each year. In a report published by Sweett and BRE (2014) a comparison of payback for three case studies (office, school and community healthcare centre) found that additional capital costs were approximately 2 per cent and payback was achieved within 2–5 years. These savings occurred through energy savings, without compromising operational performance.

3.1.7 Grants and subsidies

As previously highlighted in Chapter 2 there are a large number of grants and subsidies available. The most popular of which is the FIT which is now in place in over 40 countries worldwide, WFC (2016). In short this incentive covers photovoltaic solar electricity (Solar PV), wind turbines and hydroelectricity. The tariffs available vary depending on several factors, including when the system is installed and its capacity. Typically, there are two types of tariff (1) a generation tariff and (2) an export tariff. The former is paid for energy used on site and the later for energy that you sell back to the grid. The rates will vary depending upon the country and requirements but the majority of users will use the generation tariff as it yields a greater return. In the United Kingdom, the uptake of the FIT has been so successful that the government Department of Energy and Climate Change (DECC) have recently considered its future. As a result, the FIT review, FIT (2016) was published. In the report, the government originally proposed to cut the FIT subsidy by 87 per cent to 1.63p. However, after receiving over 55,000 written responses to their proposals they decided to instead cut the subsidy by 65 per cent from 15 January 2016. Grants and subsidies are additional key features in the incentivisation of sustainable buildings and provided the initial stimulus necessary.

3.1.8 Tax reductions

Sustainable buildings offer significant tax reductions through a number of different routes. Some of the tax incentives listed will be familiar to readers, others may not. The most common tax incentive used is Enhanced Capital Allowances (ECA), which are discussed below. Most tax incentives relate to commercial buildings and were examined by GVA Grimley Ltd (2012). They provided a useful summary of the current incentives and suggested that the following changes would stimulate sustainable commercial property construction activity:

Use of capital allowances

ECA for energy efficient assets were introduced in the United Kingdom in 2001. The purchaser of the equipment is encouraged to buy from the UK Energy Technology List (ETL), a government list of nearly 17,000 energy saving products. It is designed to encourage businesses to invest in energy saving plant or machinery. Businesses that purchase products listed on the ETL can claim a 100 per cent ECA (tax relief). GVA Grimley Ltd believes that a success and uptake has been limited and that a new rating system of energy efficiency based on the A-G system familiar with EPCs, would be more useful and would stimulate higher uptake. The plant and machinery regulations provide the current precedents for using the rating system to encourage sustainability. Further plant and machinery exemption may act as a pilot to a better rating system.

3.1.9 Increased building value

Creating a sustainable building may or may not increase the buildings value. It all depends upon the way in which the energy saving measures were funded in the design stage.

This is another reason why it is crucial to understand the financial and technical options early in design. For example, a building owner decides to retrofit a solar thermal and Ground Source Heating Pump (GSHP) system to reduce electricity and gas use. A Government Green Deal loan was arranged to fund the project and the approximate payback period is 15 years. The owner then decides to sell the property after 5 years. In this particular case an outstanding 10-year debt remains with the property. Clearly some people may not consider this to be a debt as the annual operational costs are reduced but that is a matter of opinion. If the solar thermal and GSHP systems do not perform as expected or there are insurance and warranty problems, then this is a significant debt. However, generally speaking, if the correct financial package can be arranged then energy efficiency measures (EEMs) do increase the value of a building.

3.1.10 Rental return

For commercial property, rental return is key and at each opportunity (i.e. lease renewal) it is possible to negotiate a new lease. There are a number of improvements that can be made to attract new tenants and keep existing tenants. Two of which are (1) increasing the rentable square footage of a property and (2) installing EEMs. On improvement number one, if a building is poorly designed, the rentable square feet will be significantly less than the 'actual' square feet. This means a large part of the property that could be potentially monetised is not. For example, a building with excessive lobbies, unfinished areas, or poorly organised hallway arrangements could end up having a large portion of the property not rentable. On improvement number two, the installation of new EEMs can attract a tenant as the operational costs are lower.

3.2 Health and well-being

There are a number of indirect health and well-being savings due to the creation of a sustainable building. Hakkinen and Belloni (2011), USGBC (2007), Durmus-Pedini and Ashuri (2010) and Loftness, Hartkpof, Gurtekin, Hansen, and Hitchcock (2003) found that sustainable buildings improved well-being through productivity of occupants and users of the buildings, reduced emissions and use of natural resources. A study by Leaman and Bordass (2005) concludes that offices work best for human productivity when:

1 there are opportunities for personal control;
2 there is a rapid response to environmental issues;
3 there are shallow plan forms, preferably with natural ventilation and less technical and management-intensive systems;

4 there is enough room for people to work in, and appropriate zoning and control of heating, cooling, lighting and ventilation;
5 there are occupiers who are given clear instruction of the design intent. Occupiers are shown how things are intended to work.

Support for improved Facilities Management (FM), as a means of increasing office productivity, is presented by Clements-Croome (2003). He maintains that both greater energy savings and increases in productivity can be achieved by ensuring that healthy buildings are produced. He also acknowledges that it is not just the design and construction of the building, but also the way the building is managed through its FM provision that can impact on occupier productivity. Clements-Croome, (2003) identifies that the most frequent complaints relate to thermal problems, stuffiness, sick building syndrome (SBS) and crowding. It is therefore suggested that by improving the office environmental conditions, occupier productivity could be increased by 4–10 per cent.

3.2.1 Occupier productivity

Creating a sustainable building is an incentive to occupier productivity as it is now widely accepted that occupier productivity is linked to the quality of the internal workspace. Leaman (1995) quoted that people who are unhappy with temperature, air quality, lighting and noise conditions in their offices are more likely to say that this affects their productivity at work. He also found that there is a direct correlation between people who report dissatisfaction with their indoor environment and those that report the office environment to be affecting their productivity. When constructing a sustainable building it is clearly then very important to consider the internal workspace and occupier productivity. Creating a building that can remain at an ideal temperature independent of the outside temperature, has fresh air circulation ideally through natural ventilation, is well illuminated through ideally natural lighting and results in a more content occupant. These are the traits of a sustainable building.

3.2.2 Sick building syndrome

Whitley, Makin, and Dickson (1996) proposed that occupiers' satisfaction with the indoor environment could be influenced by the climate of the organisation and the occupiers' satisfaction with their jobs. Their research aimed to investigate SBS, and its effects on occupiers, both in terms of health and productivity. They collected over 400 responses from two buildings using a psychology questionnaire with a perceived productivity scale. The productivity scale adopted, with slight modification, was the same one originally proposed by Leaman (1995) and subsequently adopted by Oseland (1999). In the study Whitley *et al.* (1996) concluded that: '*Office satisfaction is seen as a primary predictor of sick building syndrome and self-reported productivity. Office satisfaction is significantly associated with job satisfaction and environmental control.*' In an additional study by Wargocki, Wyon, Sundell,

Clausen, and Fanger (2000b) he attempted to evaluate the effects of outdoor air supply rate on perceived air quality, SBS and productivity. The evaluations were conducted in a normally furnished office. The results indicated that when ventilation was increased the subjects reported feeling generally better and had less difficulty in thinking. Therefore, the ventilation requirements of the office occupiers become an important ingredient in creating a productive workplace.

3.3 Standards and regulations

There are a large number of widely accepted standards, benchmarks and regulations to guide the development and measure the effectiveness of a sustainable building. The most established measure is the Passivhaus standard developed in Germany in the early 1990s by Professors Bo Adamson of Sweden and Wolfgang Feist, Passivhaus (2016). Other standards are used to incentivise the uptake of sustainable buildings and include for example the US Leadership in Energy and Environmental Design, LEED (2016), the UK Building Research Establishment Environmental Assessment Method, BREEAM (2016), the Australian Green Star Programme, GBCA (2016) or the Japanese Comprehensive Assessment System for Built Environment Efficiency, CASBEE (2016).

One example of a BREEAM outstanding building is One Angel Square, see Figure 3.2, an office building in Manchester (United Kingdom) which was completed in 2013. The building made use of many sustainability features including active and passive design. The building made use of natural resources, natural ventilation, a double skin façade, adiabatic cooling, rainwater harvesting, greywater and waste heat recycling.

Figure 3.2 One Angel Square

3.3.1 Benchmarks

In addition to the measures and standards identified above there are also a large number of benchmarks. These benchmarks are used to help guide investors, developers, design and construction teams and occupiers to create a more sustainable building. They provide procedures, methodologies and validation. The most widely used global benchmarks as mentioned above are BREEAM, LEED (2016), CIBSE TM46 (2016), Better Building Partnership (BBP, 2010), and the US Energy Star (2016). These benchmarks are led by building regulations which play an integral role in ensuring quality and standards.

3.3.2 Building regulations

Building regulations set minimum standards for the design and construction of buildings to ensure the safety and health for people in or about those buildings. They also include requirements to ensure that fuel and power is conserved and facilities are provided for people, including those with disabilities, to access and move around inside buildings. The main building regulations that incentivise the construction of sustainable buildings differ by country with some examples provided below:

Energy efficiency regulations in the EU

The Energy Performance of Buildings Directive (EPBD) 2010/31/EU is the main legislative instrument, at the European level, for improving the energy efficiency of buildings. A key element of the EPBD is its requirement for 'Nearly Zero-Energy Buildings (NZEB)'.

According to Article 9 of the Directive:

1. Member States shall ensure that: (a) by 31 December 2020, all new buildings are nearly zero-energy buildings; and (b) after 31 December 2018, new buildings occupied and owned by public authorities are nearly zero-energy buildings.

The first step for achieving NZEB is to have a Fabric Energy Efficiency Standard (FEES). This is calculated in kWh/m^2/year energy demand and is only achievable through ensuring a sound fabric first approach. The now redundant UK Zero Carbon Standard used this same EU NZEB approach.

Energy efficiency regulations in the United Kingdom

Building regulations come in the form of approved documents and provide general guidance on how different aspects of building design and construction can comply with the building regulations. The approved Document, Part L (2016) is concerned with the conservation of fuel and power. In addition, it deals with energy efficiency requirements and is the main regulation for the incentivisation of sustainable buildings. Part L of the UK Building Regulations 2010 is concerned with the conservation of fuel and power. There are four approved documents in the series:

- Part L1A – conservation of fuel and power in new dwellings;
- Part L1B – conservation of fuel and power in existing dwellings;
- Part L2A – conservation of fuel and power in new buildings other than dwellings;
- Part L2B – conservation of fuel and power in existing buildings other than dwellings.

Energy efficiency regulations in the United States

US building codes sets minimum requirements for energy-efficient design and construction for new and renovated buildings that impact energy use and emissions for the life of the building. There are three different code types (1) federal sector codes, (2) state sector codes and (3) local codes.

The private sector codes are associated with state and local jurisdiction. States and local jurisdictions have different energy codes that they follow based on climate, geography and many other contributing factors. The two primary baseline codes for the private sector are the International Energy Conservation Code, IECC (2012), and the ANSI/ASHRAE/IESNA Standard 90.1 energy standard for Buildings Except Low-Rise Residential Buildings ASHRAE 90.1 (2013). US states and local governments adopt and enforce these energy codes.

Energy efficiency regulations in China

China launched a green building labeling system referred to as the Three-Star Rating Building System in 2006. Under this program, buildings are rated from one to three stars according to criteria that include use of land, energy and water, in addition to material efficiency, indoor environmental quality and operational management. In addition to building design, the Three-Star Rating Building System measures a building's performance and awards a rating after 1 year of building operation.

The Chinese government has paid particular attention to retrofits and renovation of existing buildings. The purpose of this program is to bring existing buildings to the level of code required for new construction. In 2011, the government strengthened its obligation by requiring a 10 per cent reduction in energy consumption per square meter for commercial buildings and a 15 per cent reduction for large commercial buildings with more than 20,000 square meters of floor area by the end of 2015. Under the Green Building Action Plan of 2013, more than 400 million square feet in residential homes and all eligible commercial buildings in the northern heating zone are anticipated to be retrofitted by the end of 2015 and 2020, respectively, US EIA (2015).

3.4 Case studies

Sustainable buildings offer a number of benefits as noted by Edwards and Naboni (2013) and Kats (2003). They include but are not limited to:

1 energy and water savings;
2 reduced waste;
3 improved indoor environmental quality;
4 greater employee comfort/productivity;
5 reduced employee health costs;
6 lower operations and maintenance costs.

Table 3.2 below highlights the key attributes that show the increased value of a sustainable building from a large collection of US and Australian case studies. The table is reproduced from a fact sheet produced by the Institute for Building Efficiency, IBE (2016) and Pearce *et al.* (2012).

3.5 Guiding principles

The guiding principles listed in Chapter 3 are based upon the incentives to investment. They are the crucial elements for success and are listed below:

Principle 3.1 – A sustainable building can be more valuable than a standard equivalent building and the scale of that value is dependent upon the method with which the energy efficiency measures were funded in the design stage.

Principle 3.2 – Generally speaking, a sustainable building improves health and well-being through productivity of occupants and users of the building, reduced emissions and use of natural resources.

Principle 3.3 – A sustainable building ensures that current and future regulations are met and that the building remains attractive to its owners and occupants over a longer time period, thereby increasing the return on investment.

Table 3.2 Benefits of sustainable buildings

Green building benefit	*Increased market value compared with conventional buildings*	*ENERGY STAR buildings*	*LEED-certified buildings*
Increased rental rates	2–17%	ENERGY STAR properties had a rental premium of 4.8%, or $1.26 per square foot. (Pivo and Fischer, 2008) ENERGY STAR offices had a rental premium of 3% per square foot between 2004 and 2007. (Eichholtz, Kok, and Quigley, 2009) An office building registered with LEED or ENERGY STAR had a rental premium of 2% between 2007 and 2009. (Eichholtz, Kok, and Quiqley, 2010) ENERGY STAR certified office space had a rental premium of 6%. (Fuerst & McAllister, 2009a) ENERGY STAR buildings had a rental premium of 7%–9%. (Wiley, Benefield, and Johnson, 2010)	An office building registered with LEED or ENERGY STAR had a rental premium of 2% between 2007 and 2009. (Eichholtz *et al.*, 2010) LEED-certified office space had a rental premium of 5%. (Fuerst and McAllister, 2009b) LEED buildings had a rental premium of 15–17%. (Wiley *et al.*, 2010)
Improved resale value	5.8%–35%	ENERGY STAR properties had a 13.5% higher market value per square foot relative to non-ENERGY STAR properties. (Pivo and Fischer, 2008) Building sale price increased by 5.8% with ENERGY STAR certification. (Miller, Spivey, and Florance, 2008) A sale price premium of 16% was found for ENERGY STAR offices between 2004 and 2007. (Eichholtz *et al.*, 2009) A sale price premium of 13% was found for ENERGY STAR and LEED office buildings between 2007 and 2009. (Eichholtz *et al.*, 2010) A sale price premium of 31% was reported for ENERGY STAR offices. (Fuerst and McAllister, 2009a) A sale price premium of $30 per square foot for ENERGY STAR was found in 25 metropolitan markets. (Wiley *et al.*, 2010)	Building sale price increased by 10% with LEED certification in an analysis of building sales from 2003 to 2007. (Miller 2008) A sale price premium of 13% was found for LEED and ENERGY STAR office buildings between 2007 and 2009. (Eichholtz *et al.*, 2010) A sale price premium of 35% was reported for LEED-certified offices. (Fuerst and McAllister, 2009b) A sale price premium of $130 per square foot LEED-certified buildings was found in 25 metropolitan markets. (Wiley *et al.*, 2010)

Table 3.2 continued

Green building benefit	*Increased market value compared with conventional buildings*	*ENERGY STAR buildings*	*LEED-certified buildings*
Higher occupancy rates	0.9%–18%	ENERGY STAR properties had 0.9% higher occupancy rates. (Pivo and Fischer, 2008) Occupancy rates were 2%–4% higher for ENERGY STAR buildings compared with non-ENERGY STAR buildings. (Miller *et al.*, 2008) Effective rent (rent adjusted for different vacancy rates in labelled offices) was about 10% higher in ENERGY STAR offices, compared to offices of the same age and building quality, within a 0.2-square-mile area around a certified building. (Eichholtz 2009) Effective rent was 5% higher for ENERGY STAR and LEED office buildings between 2007 and 2009. (Eichholtz *et al.*, 2010) Occupancy rates were 10%–11% higher for ENERGY STAR certified buildings. (Wiley *et al.*, 2010) Occupancy rates were 3% higher in ENERGY STAR-labelled offices. (Fuerst 2009a)	Effective rent (rent adjusted for different vacancy rates in labelled offices) was about 9% higher for LEED offices, compared to offices of the same age and building quality, within a 0.2-square-mile area around a certified building. (Eichholtz 2009) Effective rent was 5% higher for LEED and ENERGY STAR office buildings between 2007 and 2009. (Eichholtz 2010) Occupancy rates were 16%–18% higher for LEED-certified buildings. (Wiley *et al.*, 2010 Occupancy rates were 8% higher in LEED-labelled offices. (Fuerst and McAllister, 2009b)
Lower operating expenses	30%	Operating expenses were 30% lower for ENERGY STAR versus non-ENERGY STAR buildings. (Miller, Spivey, and Florance, 2008) ENERGY STAR buildings had 9.8% lower utility expenditures than non-ENERGY STAR buildings. (Pivo and Fischer, 2008)	
Higher net operating income (from higher rents, higher occupancy rates, or lower operating expenses)	5.9%	For ENERGY STAR properties, net operating income per square foot was 25 cents (5.9%) higher than for non-ENERGY STAR properties. (Pivo and Fischer, 2008)	

Table 3.2 continued

Green building benefit	*Increased market value compared with conventional buildings*	*Energy efficient buildings with green attributes*	
Productivity gains rates, or lower operating expenses	4.88%	Productivity gains may increase due to: lighting (0.7%–23%), quieter working conditions (1.8%–19.8%), improved ventilation (0.6%–7.4%), and workstation controls (0.2%–3%). (Loftness *et al.*, 2003) For LEED-certified buildings, benefits of $37 to $55 per square foot are reported as a result of productivity gains from less sick time and greater worker productivity, primarily from better ventilation, lighting and general environment. (Kats, 2003) ENERGY STAR and LEED buildings had productivity gains of 4.88% for tenants who reported gains. (Miller, Pogue, Gough, and Davis, 2009)	
Australian studies			
Green building benefit	*Increased market value compared with conventional buildings*	*NABERS building*	*Green star building*
Premium in value	2%–12%	Buildings with a 5-star NABERS rating delivered a 9% premium in value, and 3–4.5 star NABERS energy ratings delivered 2%–3% premium in value. (IBE, 2016)	Green star rated buildings had a 12% premium in value. (IBE, 2016)
Investment return	0.6%–4%	The investment return on buildings with a NABERS energy rating was 0.6% higher than a nonrated building. (IPD, 2011)	The investment return was 4% higher with a Green star rating. (IPD, 2011)

References

Ala-Juusela, M., Huovila, P., Jahn, J., Nystedt, A., and Vesanen, T. (2006). *Energy Use and Greenhouse Gas Emissions from Construction and Buildings*. Final report provided by VTT for UNEP. Parts of the text published in: UNEP (2007) *Buildings and Climate Change Status, Challenges and Opportunities*, Paris, UNEP

ASHRAE 90.1. (2013). Energy standard for buildings except low-rise residential buildings. ANSI/ASHRAE/IES Standard 90.1–2013. ISSN 1041–2336

BBP. (2010). Sustainability benchmarking toolkit for commercial buildings. Building Better Partnership. Retrieved from www.betterbuildingspartnership.co.uk/sites/default/files/media/attachment/bbp-sustainability-benchmarking-toolkit.pdf (Accessed 18 April 2016)

BREEAM. (2016). Building Research Establishment Environmental Assessment Method. Retrieved from www.breeam.com/index.jsp (Accessed 18 April 2016)

CASBEE. (2016). Japanese Comprehensive Assessment System for Built Environment Efficiency. Retrieved from www.ibec.or.jp/CASBEE/english/ (Accessed 18 April 2016)

CCL. (2016). UK Government Climate Change Levy. Retrieved from www.gov.uk/green-taxes-and-reliefs/climate-change-levy (Accessed 18 April 2016)

CIBSE TM46. (2016). TM46 Energy Benchmarks. Chartered Institute of Building Services Engineers. Retrieved from www.cibse.org/knowledge/cibse-tm/tm46-energy-benchmarks (Accessed 18 April 2016)

Clements-Croome, D. (2003). Environmental quality and the productive workplace. CIBSE/ASHRAE Conference. Building Sustainability, Value and Profit. Edinburgh, Scotland (24–26 September)

DECC. (2015). Updates Energy and Emissions Projections 2015. Department of Energy and Climate Change, HM Government. Crown Copyright

DECCa. (2015). UK Energy in Brief, HM Government. Crown Copyright

Durmus-Pedini, A., and Ashuri, B. (2010). An overview of the benefits and risk factors of going green in existing buildings. *International Journal of Facility Management, 1*(1)

ECO. (2016). Office of Gas and Electricity Markets (OFGEM) Energy Company Obligation. Retrieved from www.ofgem.gov.uk/environmental-programmes/eco (Accessed 18 April 2016)

Edwards, B.W., and Naboni, E. (2013). *Green buildings pay: Design, productivity and ecology*. Abingdon, UK: Routledge

Eichholtz, P., Kok, N., and Quigley, J. (2009). Doing Well By doing Good? Green office Buildings, Working paper. Fisher Center for Real Estate and Urban Economics, UC Berkeley. Retrieved from www.ucei.berkeley.edu/PDF/seminar20090130.pdf (Accessed 18 April 2016)

Eichholtz, P., Kok, N., and Quiqley, J. (2010). The Economics of Green Building. Maastrict University and University of California – Berkeley, August 2010

FiT. (2016). UK Government Feed in tariff. Retrieved from www.gov.uk/feed-in-tariffs/overview (Accessed 18 April 2016)

Fuerst, F., and McAllister, P. (2009a). New Evidence on the Green Building Rent and Price Premium. University of Reading, 2009. Retrieved from www.henley.ac.uk/rep/fulltxt/0709.pdf

Fuerst, F., and McAllister, P. (2009b). An investigation of the effect of eco-labeling on office occupancy rates. *Journal of Sustainable Real Estate*, 1(1), 2009

GBCA. (2016). Australian Green Star Programme. Retrieved from www.gbca.org.au/green-star/ (Accessed 18 April 2016)

GSA. (2009). Benefits of Green Buildings on Costs, the Environment and Jobs. Retrieved from www.gsa.gov/portal/content/103662 (Accessed 18 April 2016)

GVA Grimley. (2012). Challenging the government on sustainability: incentivising through tax. Retrieved from www.gva.co.uk/sustainability/challenging-the-government-on-sustainability (Accessed 18 March 2016)

Hakkinen, T., and Belloni, K. (2011). Barriers and drivers for sustainable buildings. *Building Research and Information, 39*(3), 239–255, DOI: 10.1080/09613218.2011.561948

IBE. (2016). Institute for Building Efficiency. Assessing the Value of Green Buildings. Retrieved from www.institutebe.com/InstituteBE/media/Library/Resources/Green%20Buildings/Green-Building-Valuation-Fact-Sheet.pdf (Accessed 18 March 2016)

IECC. (2012). International energy conservation code. Illinois: International Code Council.

IPD Australia and New Zealand. (2011). Green Cities 2011: Introducing the PCA/IPD Green Investment Index. Retrieved from www.ipd.com/LinkClick.aspx?fileticket=e48fKnS8DKQ%3D&tabid=427&mid=10392 (Accessed 18 April 2016)

Janzen, H. and Brasok, J. (2011). Residential Property Taxes and Utility Charges Survey. City of Edmunton Planning and Development. Retrieved from www.edmonton.ca/business_economy/documents/PDF/PropertyTax_Final_2010_Final_Report.pdf (Accessed 18 April 2016)

Kats, G. (2003). Green Building Costs and Financial Benefits. Westborough, MA: Massachusetts Technology Collaborative

Leaman, A. (1995). Dissatisfaction and office productivity. *Facilities*, *13*(2), 3–19

Leaman, A., and Bordass, W. (2005). Productivity in buildings: the 'killer' variables. In Clements-Croome, D (Ed.), *Creating the Productive Workplace*, 2nd ed., London: E and FN Spon, pp. 153–180

LEED. (2016). Leadership in Energy and Environmental Design. Retrieved from www.usgbc.org/leed (Accessed 18 April 2016)

Loftness, V., Hartkpof, V., Gurtekin, B., Hansen, D., and Hitchcock, R. (2003). Linking energy to health and productivity in the built environment. Center for Building Performance and Diagnostics, Carnegie Mellon

Miller, N., Spivey, J., and Florance, A. (2008). Does Green Pay off? *Journal of Real Estate Portfolio Management*, *14*(4), Oct–Dec

Miller, N., Pogue, D., Gough, Q. and Davis, S. (2009). Green Buildings and Productivity. *The Journal of Sustainable Real Estate*, 1(1), 65–89

Olubunmi. O., Xia, P., and Skitmore, M. (2016). Green building incentives: a review. *Renewable and Sustainable Energy Reviews*, 59, pp. 1611–1621

Oseland, N. (1999). *Environmental factors affecting office worker performance: A review of evidence.* Technical Memoranda TM24. London: CIBSE

Part, L. (2016). Conservation of fuel and power: Approved document L. Retrieved from www.gov.uk/government/publications/conservation-of-fuel-and-power-approved-document-l (Accessed 18 April 2016)

Passivhaus. (2016). The Passivhaus Standard. Retrieved from www.passivhaus.org.uk/ (Accessed 18 April 2016)

Pearce, A.R., Ahm, Y.H., and Hanmi Global. (2012). *Sustainable buildings and infrastructure: Paths to the future.* Abingdon, UK: Routledge

Pitt, M., Tucker, M., Riley, M., and Longden, J. (2009). Towards sustainable construction: Promotion and best practices. *Construction Innovation*, *9*, 201–224

Pivo, G., and Fischer, J. (2008). Investment Returns from Responsible Property Investments: Energy Efficient, Transit-oriented, and Urban Regeneration Office Properties in the US from 1998–2007. Indiana University

RO. (2015). The Renewables Obligation for 2016/17. UK Department of Energy and Climate Change. Crown Copyright 2015. URN 15D/461

Sweett and BRE. (2014). Delivering sustainable buildings: savings and payback. Berkshire, UK: BREPress

UK House of Commons. (2008). House of Commons Business and Enterprise Committee Construction Matters: Ninth Report of Session 2007–2008, 2, p. 189

UNEP. (2010). Green Buildings and the Finance Sector. CEO briefing: A document of the UNEP FI North American Task Force. Retrieved from www.unepifo.org

US EIA. (2015). Chinese policies aim to increase energy efficiency in buildings. US Energy Information Administration. Retrieved from www.eia.gov/todayinenergy/detail.cfm?id=23492 (Accessed 18 April 2016)

US Energy Star. (2016). US Energy Star Programme. Retrieved from www.energystar.gov/ (Accessed 18 April 2016)

USGBC. (2007). A National Green Building Research Agenda. Retrieved from www.usgbc.org/ShowFile.aspx?DocumentID=3402 (Accessed 18 April 2016)

USGBC. (2009a). Green buildings supporting US jobs. Retrieved from www.usgbc.org/Docs/News/Green%20building,%20green%20jobs%20and%20the%20economy%20-%20Booz%20Allen%20report%20GS.pdf (Accessed 18 April 2016)

USGBC. (2013). The Business Case for Green Buildings: A review of the costs and benefits for developers, investors and occupants. World Green Building Council. Retrieved from www.worldgbc.org/files/1513/6608/0674/Business_Case_For_Green_Building_Report_WEB_2013-04-11.pdf (Accessed 12 June 2016)

Wargocki, P., Wyon, D. P., Sundell, J., Clausen, G., and Fanger, P.O. (2000b). The effects of outdoor air supply rate in an office on perceived air quality, Sick Building Syndrome (SBS) symptoms and productivity. *Indoor Air, 10,* 222–236

WFC. (2016). Feed In Tariffs – A guide to one of the world's best environmental policies. World Future Council. Retrieved from http://area-net.org/wp-content/uploads/2016/01/WFC_Feed-in_Tariffs_Brochure.pdf (Accessed 12 June 2016)

WHD. (2016). UK Government Warm Home Discount. Retrieved from www.gov.uk/the-warm-home-discount-scheme/what-youll-get (Accessed 18 April 2016)

Whitley, T. D. R., Makin, P. J. and Dickson, D. J. (1996) 'Job Satisfaction and Locus of Control: Impact on Sick Building Syndrome and Self-Reported Productivity', 7th International Conference on Indoor Air Quality and Climate, Nagoya, Japan

Wiley, J., Benefield, J. and Johnson, K. (2010). Green design and the market for commercial office space. *Real Estate Finance and Economics, 41*(2)

4 Barriers to investment

In Chapter 3, the incentives to investment in sustainable buildings were discussed. The various and numerous options are well established and researched; demonstrating how sustainable buildings can benefit the owner and occupier. In this chapter, the focus is switched to the barriers to investment and you may well consider the only significant barrier to investment in sustainable buildings as being cost and cost alone. However, this is untrue as evidenced by Hakkinen and Belloni (2011). The main barriers are in fact cost, steering mechanisms, client understanding, process, procurement, tendering, timing, cooperation, knowledge, availability of methods and innovation. The details of which are explained in this chapter.

Continued innovation in technology

Consider for a moment the recent advancements in mobile phones. The world's first mobile phone was created on the 3 April 1973 by Motorola weighing 1.1 kg with 30 minutes of talk time, requiring a 10-hour charge. Now consider the use of building energy efficiency measures (EEMs) such as solar photovoltaic (PV) panels. At present there are 3 main classes of PV cell with over 300 different manufacturers. What if you were to invest in a solar PV array only to discover that 6 months later there is a new type of system that is twice as efficient at half the price! It is for this and many other reasons that continued innovation in technology causes individuals to be concerned. This is why many individuals, who wish to source EEMs for buildings, will contract a third-party provider to carry the technical risk (as discussed in Section 2.3.3 with the use of Energy Services Companies [ESCOs]). Technology will always advance but at some point there is a need to make a commitment to change and if the client decides to engage with a third-party provider it is in their best interest to adopt the very latest innovation as that will be beneficial for all. As a result, they would replace the equipment as and when required as part of the ongoing contractual arrangement.

Commitment to the production of a sustainable building does require action and decisions to be made. This is why it is wise to undertake a review of options and make a decision based upon a number of techniques such as: Best Available

Technology Not Entailing Excessive Cost (BATNEEC) approach, Life Cycle Assessment (LCA) and LCCA as discussed by Kulczycka and Smol (2016). Using these methods will ensure that those involved select the correct technical option. Prior to selection of options there is of course a requirement to commit to change.

Commitment to change

The construction of a new building or retrofit of an existing building is process driven through either a design/bid/build or design/build or retrofit strategy respectively. The processes for new buildings are well established, following government guidelines and prescribed methods such as the Royal Institute of British Architects (RIBA) Plan of Works, which originated in 2007 and has been updated in 2013, RIBA (2013). The plan of work organises the process of briefing, designing, constructing, maintaining, operating and using building projects into a number of key stages. The content of stages may vary or overlap to suit specific project requirements. It is used solely as guidance for the preparation of detailed professional services contracts and building contracts. The RIBA recognised that this procedural approach acted as a barrier to investment into sustainable buildings and therefore committed to changing the approach. As such they have created the '*Green Overlay*'. This overlay provides sustainability checkpoints against each of the RIBA work stages to aid in the creation of a sustainable building. Each checkpoint is supported by supplementary guidance to guide the user as necessary, RIBA green overlay (2011). For the retrofit of existing buildings, there are less-established methodologies and frameworks. Useful methodologies have been created by Sturgis Carbon Profiling, Sturgis (2016).

If you are considering sustainable design for a new building using the RIBA green overlay or the refurbishment of an existing building using the techniques prescribed by Sturgis; you have made that commitment to change. This is the starting point for the creation of a sustainable building. One of the next major barriers to investment is to consider investment and raising the necessary capital.

4.1 Capital expenditure

In terms of direct capital expenditure, sustainable buildings often involve higher initial investment costs than conventional building. The Sustainable Construction and Innovation Group, SCI (2011) states that these higher construction costs typically arise due to:

- more expensive design work, for example innovative heating, cooling or renewable energy systems;
- higher material costs, for example to apply a better standard of insulation or to include solar installations;
- costs associated with third-party certification to ensure compliance with an environmental standard i.e. Building Research Establishment Environmental Assessment Method (BREEAM);

- costs associated with minimising the environmental impact of the construction process, for example waste and resource management, noise reduction, minimising impact on flora and fauna;
- specific contractual provisions relating to intellectual property, insurance and indemnities, working conditions or other factors, which may result in higher costs being passed to contractors.

Conversely, however building with high environmental performance will lead to lower whole-life costs. For example, Sydney's One Central Park, see Figure 4.1, has embraced environmental design through the use of water recycling, low-carbon tri-generation power plant and the use of hydroponic vertical gardens. As a result of the initial expenditure the buildings energy use is significantly reduced.

The balance between the initial investment and return on investment needs to be carefully considered and it is for this reason that sustainable buildings should be considered using LCCA and not capital expenditure only. Most developers and building owners still consider the capital expenditure over LCCA and again this acts as a further barrier to investment.

4.1.1 Business case for investment

When considering the amount of capital expenditure required for traditional and EEMs for new buildings or the retrofit of existing buildings; it is necessary to understand the business case for investment and who the beneficiary is. If the business case is not outlined correctly these proposed measures act as a barrier to investment. For a residential or commercial development, the beneficiary could be the owner/developer/landlord/tenant or occupant. For each party there

Figure 4.1 Sydney's One Central Park

is a different incentive. For example, a home owner may wish to reduce their energy bills by purchasing a more energy efficient boiler, the commercial property developer knows he/she will be able to command a higher rent or better lease arrangement for a more sustainable building. The tenant may be attracted to a more sustainable building as it (1) results in cheaper running costs and (2) is good for their corporate image. The building owners/developers will wish to create a more sustainable building that yields a return on investment. That return could be financial, social and/or environmental. In simplistic terms, outlining the business case for the investment into EEMs requires a knowledge of three key considerations:

1 the initial capital expenditure (CAPEX);
2 the operational expenditure (OPEX);
3 Return on Investment (ROI).

In addition, there are clearly a number of other key considerations such as flexibility of system, energy saving potential and technical feasibility. For all building types an outline and full business case for all options is required. Individuals will seek to make a judgement as to whether the project is affordable in line with other objectives. In forming this judgement, teams will need to consider the balance of risks associated with the project. There should be an iterative process in order to develop satisfactory proposals but it is also permitted to reject a business case that you believe represents a poor strategic option, is unaffordable or represents poor value for money. In short, the business case could cover the:

(a) Strategic Outline Case (SOC) – making a case for change and providing an early indication of options through a SWOT analysis.
(b) Outline Business Case (OBC) – which revisits the SOC in more detail and identifies preferred options which demonstrate Value for Money (VfM).
(c) Full Business Case (FBC) – which takes place in the procurement stage of services following detailed negotiations with potential service providers and/or suppliers.

Developing a business case for investment using these three stages greatly assists in decision making and reduces the perceived risk. However, sometimes the incentives are split between the beneficiaries, making the business case slightly more difficult to assess.

4.2 Split incentives

Split incentives are generally a barrier to the deployment of EEMs in buildings. Split incentives occur when those responsible for paying energy bills (the tenant) are not the same entity as those making the capital investment decisions i.e. the landlord or building owner. They are therefore only applicable to this specific type of development. In these circumstances, the landlord may not be inclined to make the necessary upgrades to building services when the benefits associated with the resulting energy savings accrue to the tenant.

A key mechanism for overcoming split incentives has been investigated by the Australian Government Department of Industry, Innovation and Science (AGDIIS) who produced a fact sheet, AGDIIS (2013). Their solution is to use '*Green Lease*' clauses to the lease agreement between the landlord and tenant which typically include a:

- commitment by the landlord to maintain standards;
- commitment by the tenant to achieve and maintain an energy management plan;
- establishment of a building management committee that meets periodically to discuss ways to improve the energy efficiency and sustainability of the building;
- dispute resolution process.

Green leases can overcome split incentives by setting mutually agreed performance targets and defining the actions each party will take to uphold their side of the agreement, and encouraging dialogue and collaboration between the landlord and tenants to achieve those shared goals. The owner/occupier relationships are therefore key.

4.2.1 Owner/occupier relationships

The relationship between the owner and occupier of a building has a crucial role in the creation of sustainable buildings and can act as a significant barrier. In 2000, David Cadman introduced the concept of the 'circle of blame', see Figure 4.2. Investors, occupants, contractors and developers blame each other for the lack of sustainable buildings, Keeping (2000).

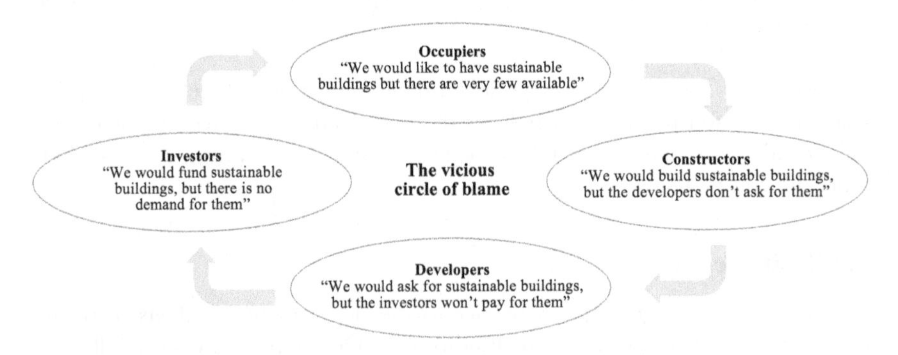

Figure 4.2 Circle of blame

Reproduced from Keeping (2000)

For example, property investors think that even if they wanted to invest in sustainable buildings, tenants would not want to lease them. In contrast, tenants might think that even if they wanted to live in a sustainable building, there is only a limited supply of them on the market. In addition, the opinion of developers is that investors are not interested in sustainability and because developers do not ask for sustainable buildings, constructors are not willing to build them. This thinking creates an endless circle. The vicious circle of blame could be resolved without difficulty if all market players would simultaneously change their views from negative to positive, Andelin (2015). A study by Cajias, Geiger, and Bienert (2012) examined how the vicious circle of blame could be reversed and made into a virtuous circle of sustainability adaptation. The focus of the study is on the investor-tenant relationship because they are the end-clients in the real estate supply chain, and simultaneously, the majority of environmental impacts and emissions are caused during the use phase of a building. Therefore, the investor-tenant relationship will play a key role in reversing the cycle of blame. Hence, the objective is to identify the investor and tenant drivers for creating sustainable real estate and to discover the possible mutual interests that would provide the impetus to reverse the circle of blame with respect to sustainable buildings, as identified in studies by Eichholtz, Kok, and Quigley (2012) and Wiencke (2012). In addition, Andelin *et al.* (2015) stated that it still remains unclear whether green buildings have a cost premium and whether tenants are willing to pay premiums for leasing sustainable. This split incentive problem is confounded by the lack of incentives by governments to break the circle of blame.

4.3 Lack of incentives

Are governments really doing enough to support the creation of sustainable buildings and break the circle of blame? Is there enough interest from the public and private sector to want to live and work in sustainable buildings? There is mixed opinion on the answers to these questions and in the recent work undertaken by Sun, Geelhoed, Caleb-Solly, and Morrell (2015) the answer was unequivocally no. In their study on the knowledge and attitudes of small builders in the United Kingdom, a series of questions were asked to over 100 professionals with the vast majority confirming that there are too few incentives for sustainable buildings. The biggest most important factor as highlighted in a further study by Sourani and Sohail (2011) is the lack of funding and reluctance to incur higher capital costs. In 75 per cent of the people interviewed in their study, this was the largest barrier. In a recent energy survey of commercial buildings by Tuffin Ferraby Taylor, TFT (2016) they found that 80 per cent of commercial property landlords claimed that the lack of government incentives is the single biggest barrier to widespread EEMs being introduced across the UK commercial real estate sector. Clearly then, in order for sustainable buildings to become mainstream there is a need to provide the right series of incentives. The incentives for existing buildings will be different from that of new buildings. Furthermore, the incentive for residential buildings will differ from those for non-residential buildings. There is no one fit solution and the incentives should be tailored accordingly. Clearly there are some general measures and technologies that will work on all buildings such as insulation. Other more bespoke measures are of equal importance and the right financial and supporting long-term mechanisms need to be in place. The EEMs have to work on a cost neutral basis and this is achievable, as discussed in Chapter 2, through the use of an ESCO. Moreover, as technology advances the capital cost of equipment will decrease and the technical options will become more viable without the need for government incentives and support such as the Feed in Tariff (FIT) or Renewables Obligation Certificates (ROCs). However there remains a lack of support for all sustainability measures.

4.3.1 Support for sustainability measures

Sustainability measures such as solar PV and other systems have received government support in the majority of countries, but not to the extent necessary. Generally, the support is financial through some form of tariff that supports the export or self-use of energy. For example, the FIT can be used when a user generates their own electricity. They can benefit from a *generation* or *export* tariff. The *generation* tariff will earn the user an income for every kilowatt hour (kWh) of electricity they generate which is used in the property. The *export* tariff can be used for every kWh of electricity that is sold to the grid. The tariff for each is different depending upon a number of factors and historically has been fixed over a long period of time. The FIT has been used by numerous countries as a method to ensure security of supply and stimulate the market to create sustainable buildings. However once a country reaches a critical mass and/or the amount of money available for the subsidy runs out, then it will cease to exist. Unfortunately, this has been a problem in many countries. For example, Italy has decided to halt solar subsidies following a period of investment to stimulate uptake, under the *Conto energia and Salva Alcoa law*, Mahalingam and Reiner (2016). The Italian government now believes that there is no longer a need to offer support and the United Kingdom has drastically reduced the amount of funding available.

Support for sustainability measures is essential from the onset but as the technology matures and the uptake increases, they can become self-sustaining and as a result financial incentives are reduced. This maturity occurs when there is enough demand for the products.

4.3.2 Lack of demand

A further barrier to investment is the demand for EEMs. For a residential or non-residential building any EEMs beyond those that are required for building regulations i.e. insulation, are generally not considered. But, why is this the case? It is simply because they are perceived as an additional upfront capital expenditure that does not provide an immediate return on investment. This is of course untrue and is a perceived problem, however a large number of studies from across the world, Samari, Godrati, Esmaeilifar, Parnaz, and Wira Mohd Shafiei (2013),

Durmus-Pedini and Ashuri (2010), IPF (2009), Kasai and Jabbour (2014) and EPF (2016) have found that there remains a lack of demand for sustainable building and technologies. This is primarily due to the lack of awareness, training and education, perceived problems, higher costs, new rules and regulations and lack of similar demonstrators. In addition to the lack of demand is the lack of appropriate labelling and standards. What real incentives do the current labelling systems provide? Are the energy and carbon efficiency standards sufficient to stimulate demand?

4.3.3 Lack of labelling/standards

Labelling, standards and benchmarks have an important role to play in the creation of a sustainable building and at present they are insufficient and act as a barrier to investment. This statement is supported by the work of Sourani and Sohail (2011) who have highlighted that insufficient and confusing guidance and standards are other significant barriers to the creation of sustainable buildings. Although there are tools and standards in place such as Leadership in Energy and Environmental Design (LEED) in the United States, BREEAM in the United Kingdom and Europe or the Australian Green Star Programme; there remains a lack of clarity as to how, when and why they should be used. For example, Sydney Opera House, see Figure 4.3, was awarded a 4-star Australian Green Star rating through the use of its historic seawater cooling system and use of chilled ceilings through to the more modern refurbishment of energy efficient LED lighting and use of temperature control in different ones of the building. This is

Figure 4.3 Sydney Opera House

of course commendable but what real commercial value does the label itself bring? Would the facilities management team have introduced the measures anyway?

As a result, there is a growing need to develop simple but comprehensive tools and techniques. Labels, standards and benchmarks have a role to play but that role is dependent upon the aspirations and requirements of the client and the building. For example, the client may want to make a bold statement and show people the building is a sustainable one by introducing a series of wind turbines, a green roof and/or solar PV. Clearly, having this technology does not automatically make the building sustainable but the perception is good. Alternatively does the client want the technology hidden within the building in a subtle way, relying on certification to prove the building is sustainable? This is a matter of opinion and it is clear that new technology and advancements in existing building services technology have a crucial role to play.

4.4 New technology

There are new advances in energy efficient technology for buildings on a monthly basis. From advanced solar panel glazing and 'smart glass' to cool roofs which reflect sunlight and heat to keep the building cooler. One of the barriers to the uptake is the complexity and constant change in innovation. Common questions arise such as: What happens if I choose this technology today and a better system comes along in 6 months-time? What if this technology does not integrate with my existing building? What happens if there is a problem? All of these questions and many others are of course very important and need to be addressed. The author of this book has created a new method of appraisal to assist with decision making in this area, Finnegan *et al.* (2015). The main concern with new technology is risk – not only risk that the systems will break down or will be too costly, but other unintended consequences. Mullich, J (2013) published an article for the *Wall Street Journal* following an interview with a senior vice president of a large mutual insurance company who commented *'the concern is from the unintended consequences of using green technologies, and that falls into different buckets such as construction materials and the different ways to produce and save energy'.*

The construction industry itself has a general high rate of inherent risk, with many projects failing to meet deadlines, cost and quality targets. In many cases the risk of time and cost overruns can and does compromise the economic viability of the project, making a potentially profitable investment untenable. Compared to many other activities, construction is subject to more risks due to unique features such as long duration, complicated processes, unpredictable environment, financial intensity and dynamic organisation structures, Designing Buildings (2015). Now add to the equation, EEMs and/or sustainable technology and anyone can see why this could be a barrier to investment as the perceived risk is increased. The answer to avoiding this risk has been discussed in Chapter 2 and is all dependent upon the method of finance chosen, however there is still a requirement for built environment professionals involved in each stage of the construction and/or refurbishment of a building to understand the basic systems and concepts.

Electing to involve a third-party ESCO will of course reduces the risk to zero and should be considered the best option.

4.4.1 Understanding the systems

Whether or not the client decides to involve a third-party ESCO, it is necessary to understand the systems and resultant cost implications. Key to choosing the right new technology or EEM is understanding how the various and numerous existing building services work. It is not necessary to undertake an extensive study into the intricacies of each service but it is necessary to understand the basics. Without this knowledge, new EEMs can act as a barrier to investment as the cost savings are unknown. If a building owner is educated into how these measures work and is comfortable with how they perform, they are likely to advocate their use. For example, the owner must appreciate and understand how the building is powered, heated and cooled. Knowledge of how such systems work provides the decision maker with the confidence to make the right choice and consider the practical and financial options available. With regards to electricity, the majority of buildings are powered from the national grid. If this source can be replaced by an alternative low or zero carbon source that will greatly assist in creating a sustainable building. Solar PVs are one method, others include algae as can be seen on the Bio Intelligent Quotient (BIQ) building, see Figure 4.4, in Hamburg, Germany whereby a bioreactor façade is used to produce electricity and control lighting, DesignBoom (2013).

Figure 4.4 Bio Intelligent Quotient (BIQ) building

Alongside electricity for power, there are systems that are used to either heat or cool a building. For heating it is useful to consider the following: heat sources, generation, distribution, delivery, transfer and control. For cooling it is necessary to consider both passive and active cooling. A brief account is given below from Designing Buildings (2016) and is a useful starting point for understanding how the various and numerous systems work and can therefore act as a barrier to the investment in a sustainable building.

Heat sources: There are numerous sources from which to obtain heat: solid fuel such as timber or coal; liquid such as oil or liquefied petroleum gas (LPG); gas from either natural gas or biogas; electricity from the grid, wind turbines, hydroelectricity or PVs; water from solar thermal, geothermal, ground or water source; air source; heat recovery, passive through solar gain or high thermal mass and internal heat load through people and equipment.

Heat generators: Heat can be generated by boilers, fuel burners, combined heat and power (CHP) plants, electric and gas heaters or heat pumps.

Heat distribution: Once heat is generated it can be distributed. Heat generators can be local, centralised or distributed either within a single building or on a wider area as part of a district heating scheme. Heat itself can be distributed by air blown through ducts, plenums or occupied spaces; water pumped through pipework; steam distributed through pipework; passive air movement or passive diffusion of heat through thermal mass.

Heat delivery: Distributed heat can be delivered within a space by: fan coil units; air handling units; radiating panels or embedded pipes in thermal mass.

Heat transfer: Delivering heat is through four methods: radiation, convection, conduction or phase change.

Controls: Control of heat is then required with through; manual or automated thermostats, switches or dampers; centrally by manual or automated thermostats, switches or dampers or Building Management Systems (BMS).

Passive Cooling: Passive or 'natural cooling' can be provided by: natural ventilation, thermal mass, evaporative cooling or devices such as shading, insulation or green roofs.

Active Cooling: Active cooling can be provided by: heat exchangers, mechanical or forced ventilation, chilled water, refrigerants or evaporative cooling. Active cooling can also be provided as part of a Heating, Ventilation and Air Conditioning (HVAC) system.

With knowledge of the above systems it is now possible to consider the role of sustainable technology and how they can integrate with the different systems in either a heating or cooling function. For example, could a micro wind turbine be used to generate electricity that could then be used to heat a building? If so, what size of turbine is needed and how much electricity could be generated? How much would the equipment cost and what would be the return on investment? Understanding how the existing systems and new systems could integrate is key to avoid EEMs being a barrier to investment. A full list of EEMs has already been presented in Table 1.2 in Chapter 1. Once a decision has been made to acquire this new equipment the next key step is to consider procurement and tendering.

4.4.2 Procurement and tendering

A further barrier to the investment and creation of a sustainable building is procurement and tendering. Consider a standard construction project and the existing procurement and tendering complications. There are numerous types of tendering methods, including open, selective, negotiated, serial, framework single-stage and two-stage and public procurement. A summary of each is provided in the 2016 Designing Buildings wiki, Designing Buildings (2016). As highlighted in the wiki, each tender has specific arrangements and procedures. For example, on public projects or projects that include a publicly funded element it may be necessary to advertise contracts. This is a requirement of the Public Contracts Regulations, PCR (2015) which are intended to open up public procurement within the European Union and to ensure the free movement of supplies, services and works. For example, the PCR in the European Union require all new procurement to be undertaken through the Official Journal of the European Union, OJEU (2016). Consider then any new EEMs and the requirements for procurement and tendering on both private and publicly funded projects. It is not a straightforward process and is project specific. A paper presented at the 2005 Action for Sustainability Conference in Tokyo investigated the procurement and process design for sustainable buildings, Boswell and Walker (2005). They concluded that fundamental concerns in procurement are hindering progress towards the construction of sustainable buildings, with procurement acting as a significant barrier to uptake. In order to build sustainability into tendering the London Borough of Camden developed a toolkit to provide guidance. This could be a useful starting point for developing further tenders for the use of EEMs on sustainable buildings. The toolkit covers risk assessments and their specific impact on the triple bottom line of environmental, social and economic aspects, LCPE (2016). In addition to tenders, one must also carefully consider sustainable procurement and another set of useful measures are the UK Government Buying Standards (GBS). These standards can be used to ensure that organisations meet their needs for goods, services, works and utilities in a way that benefits not only the organisation, but also society and the economy, while minimising damage to the environment, Defra (2015). A further barrier to investment in the procurement of EEMs is the change in government structure or policy. A recent example is the United Kingdom's exit from the European Union. When the United Kingdom leaves the European Union will they be required to procure goods and services through OJEU? Will it become easier (or more complex) to procure goods and services with less or more bureaucracy?

4.5 Changing government policies

There is universal acceptance that changing government strategy and policy stifle the development of sustainable buildings. The lack of a long-term robust plan creates confusion and uncertainty at a time when these two factors are of crucial importance. In 2009 the United Nations Environment Programme UNEP (2009), produced a buildings and climate change summary for decision makers prior to the 15th Conference of Parties meeting (COP15). In this publication they discussed and created the eight key building blocks that are needed for governments to successfully create sustainable buildings. These building blocks commented on the need for governments to take the lead on prioritising the building sector in their national climate change strategies and set long-term achievable targets. Without long-term government policy and supporting incentives such as the FIT and many others the business case does not stack up and progress is restricted. Governments themselves will not support (or seem to favour) a particular type of ESCO as they are independent and private sector funded. As highlighted in section 2.2.3 and reproduced below there are four basic types of ESCO:

1. *Independent ESCOs* – these are ESCOs not owned by an electric or gas company, an equipment manufacturers or an energy supply company.
2. *Building equipment manufacturers* – these are ESCOs owned by building equipment or controls manufacturers. Many will have parent companies and act as private companies.
3. *Utility companies* – these are ESCOs owned by regulated or non-regulated electric or gas companies.
4. *Energy or engineering companies* – these are ESCOs owned by international oil companies, non-regulated energy suppliers or large engineering firms.

Governments carefully consider how public funds are used and in an attempt to contend with the private sector ESCO market, the UK government established the Green Deal, which was based upon the principles of an Energy Performance Contract (EPC). Unfortunately, the project was not successful and was scrapped in 2015 for a number of reasons as discussed in this book. One of which was the complexity and public understanding of how an EPC works. Since this time, there have been additional changes in UK government policy which have led to uncertainty in the market for sustainable buildings. For example, the 2007 UK Labour government's 'building a greener future' policy confirmed that all new homes would be zero carbon by 2016. This was scrapped in 2015. Similarly, the Chinese government has no policy on the creation of zero carbon buildings in its latest 13th 5-year plan (2016–2020). However, they have announced plans to invest US$6.6 trillion in the coming decade on low carbon technology, renewables and energy efficiency. Furthermore, there is a goal of having green buildings that account for 30 per cent of new construction projects by 2020, ChinaDaily (2016). Long-term fixed policies are needed in order for the correct long term incentives and tariffs to be successful.

4.5.1 Changes to incentives and tariffs

A significant barrier to the investment in sustainable buildings is the lack of long-term incentives or indeed changes to existing incentives. One example of this disparity is the way in which the UK government introduced and managed the

FIT. The FIT was first introduced on the 1 April 2010 under the Energy Act 2008, HM Government (2008). In 2010, the UK Treasury set an annual budget covering the years 2010 to 2015. Any overspend from year one would have to come from year two as the budget was fixed. The starting FIT generation rate was set at 41.3p/kWh for a period of 25 years and returned such a good return that the uptake was much greater than was expected. As a result, the amount of monies available quickly diminished and the rate was reduced. The budget for the year 2011–2012 was originally set at £80 million but uptake was £89.6 million. On 1 August 2012, the FIT rate began at 16p/kWh and the length of the scheme was reduced to 20 years. On 1 October 2015 the rate reduced further to 12.47p/kWh, FIT (2016). The early adopters will continue to benefit from the initial 41.3p/kWh FiT rate until 2035, whereas more recent entrants receive approximately 60 per cent less. This downward amending for installations in subsequent years is typical and examples can be seen in Fell (2009), Klein *et al.*, 2010, Couture, Cory, Kreycik, and Williams (2010) in Germany and the United States. Due to the nature in which this funding became depleted, the current FIT does not currently provide the level of investment necessary. Alizamir, de Véricourt, and Sun (2016) commented on the fact that the success of FIT schemes critically depends on the tariffs set by governments which, in turn, determines the level of profitability for investors, Mendonça, Jacobs, and Sovacool (2009), Fell (2009). Overall, as has happened in the United Kingdom, over generous tariffs with higher levels of profitability attract a wider range of investors who consume the available funds, Lange (2012). This is generally at a cost to taxpayers as the incentive is not used for its original purpose, which is of course to ensure more sustainable buildings. Longer term, sustainable buildings will need to become economically viable without incentives and the only true to method to engender mass uptake is through small annual changes in building regulations which force individuals to make improvements.

4.5.2 Building regulations

Building regulations are a list of requirements that ensure minimum standards for health, safety, welfare, energy efficiency and sustainability. Most countries adopt building regulations and they are key to ensuring the long-term sustainability of new and existing buildings. At present a number the regulations can act as a barrier to investment in sustainable buildings. Two specific examples of why this is the case are presented below.

United Kingdom

Buildings within the United Kingdom are regulated under the 2010 building regulations through a series of approved documents. These documents run from A to Q and include for example structure, fire safety, ventilation and of importance to this book fuel and power, HM Government (2016). The approved Document L is concerned with the conservation of fuel and power in dwelling and non-dwellings. There are four documents in total and each sets the minimum standards necessary to construct a new building or amend an existing building. The documents have design and quality standards for the use and conservation of fuel and power in all buildings. They have been updated in 2011, 2013 and 2016 and on each occasion the design criteria has changed with developers required to make further improvements. This is a gentle method of ensuring that buildings are more sustainable; however, some would argue that the regulations are insufficient in providing the types of buildings that we need for long-term sustainability and as a result they act as a barrier to uptake.

China

A review of the use of building regulations has been undertaken by Geng, Dong, and Xue (2012) in China. There are several Chinese standards for green buildings as explained by GOBAS-Group (2003), MOHURD and GAQSIQ (2006) and MEP (2007). To improve the overall eco-efficiency of Chinese buildings using building regulations, China began to design a green building evaluation system in early 2000s. In 2001, China's Eco-house technical evaluation handbook was prepared (Li, 2010). Later, three upgraded editions were released in 2002, 2003 and 2007. The most recent edition, namely the fourth edition of China's Eco-house technical evaluation handbook, consists of two chapters and four appendices (Nie, Qin, Jiang, Zhang, and Cai 2007). To further promote the development of green buildings, the Ministry of Housing and Urban–Rural Development (MOHURD, former Ministry of Construction) released a national green building evaluation standard. This standard, nicknamed as 'three-star standard', is the first national standard on green building and became effective on 1 June 2006. It considers all the stages of one building's life cycle and covers both residential and public buildings (including office buildings, mall buildings and hotel buildings). The aim is to reduce total resource, water, energy and land use for one building and address the following six aspects, namely, land saving and outdoor environment, energy saving and utilization, water saving and utilization, material saving and utilization, indoor environment, operation and management. As with UK building regulations, these standards are positive but insufficient to ensure long term sustainability and act as a barrier.

4.6 Guiding principles

The guiding principles for the barriers to investment in sustainable buildings are listed below:

Principle 4.1 – Sustainable buildings require a commitment to change and from a cost perspective should be considered using Life Cycle Costing Analysis (LCCA) and not capital expenditure only.

Principle 4.2 – The incentive to build a sustainable building is dependent upon who the beneficiary is. Sometimes the incentives are split and the relationship between the owner and occupier is crucial.

Principle 4.3 – The current incentives and standards are ineffective in encouraging the widespread introduction of sustainable buildings. There is a growing need to develop simple but comprehensive assessment tools.

Principle 4.4 – Installing new EEMs will require a basic understanding of existing building services and the provision of new procurement and tendering procedures.

Principle 4.5 – Changing government policy and current building regulations are acting as a barrier to the uptake of sustainable buildings.

References

AGDIIS. (2016). Green Leases. Retrieved from www.environment.gov.au/system/files/energy/files/hvac-factsheet-green-leases.pdf (Accessed 15 December 2016)

Alizamir, S., de Véricourt, F., and Sun, P. (2016). Efficient feed-in-tariff policies for renewable energy technologies. *Operations Research, 64*(1), 52–66

Andelin, M., Sarasoja, A., Ventovuori, T., and Junnila, S. (2015). Breaking the circle of blame for sustainable buildings – evidence from Nordic countries. *Journal of Corporate Real Estate, 17*(1), 26–45

Boswell, P., and Walker, L. (2005). Procurement and process design. *Action for Sustainability World Sustainable Building Conference,* Tokyo, 27–29 September 2005. Issue Paper 15

Cajias, M., Geiger, P., and Bienert, S. (2012). Green agenda and green performance: empirical evidence for real estate. *Journal of European Real Estate Research, 5*(2), 135–155

ChinaDaily. (2016). China to boot construction of green buildings. Retrieved from http://usa.chinadaily.com.cn/business/2012–05/07/content_15224053.htm (Accessed 2 July 2016)

Couture, T., Cory, K., Kreycik, C., and Williams, E. (2010). A policymaker's guide to feed-in tariff policy design. Technical report, National Renewable Energy Laboratory, Washington, D.C Retrieved from www.nrel.gov/docs/fy10osti/44849.pdf (Accessed 22 July 2016)

Defra. (2015). Sustainable Procurement: the GBS for Construction Projects and Buildings. UK Department for Environment, Food and Rural Affairs

DesignBoom. (2013). World's first algae powered building by splitterwerk. Retrieved from www.designboom.com/art/worlds-first-algae-powered-building-by-splitterwek/ (Accessed 2 July 2016)

Designing Buildings. (2015). Risk in building design and construction. Retrieved from www.designingbuildings.co.uk/wiki/Risk_in_building_design_and_construction (Accessed 2 July 2016)

Designing Buildings (2016) Negotiated tendering. Retrieved from www.designingbuildings.co.uk/wiki/Negotiated_tendering (Accessed 9 July 2016)

Durmus-Pedini, A., and Ashuri, B. (2010). An overview of the benefits and risk factors of going green in existing buildings. *International Journal of Facility Management, 1*(1)

Eichholtz, P. Kok, N., and Quigley, J. (2012). The economics of green building. *Review of Economics and Statistics, 95*(1), 50–63

EPF (2016) Promoting sustainable and inclusive growth in emerging economies: Green Buildings. Economic Policy Forum. Retrieved from https://economic-policy-forum.org/wp-content/uploads/2016/02/Sustainable-and-Inclusive-Growth-Green-Buildings.pdf (Accessed 1 July 2016)

Fell, H. (2009). Feed-in tariff for renewable energies: an effective stimulus package without new public borrowing. Member of the German Parliament, Berlin, Germany

Finnegan, S., dos-Santos, J., Chow, D., Yan, Q., and Moncaster, A. (2015). Financing energy efficiency measures in buildings – a new method of appraisal. *International Journal of Sustainable Building Technology and Urban Development, 6*(2), 62–70

FiT (2016) Feed in Tariffs. Information on payments for renewable electricity in the UK. Retrieved from www.fitariffs.co.uk/FITs/ (Accessed 23 July 2016)

Geng, Y., Dong, H., and Xue, B. (2012). An overview of chinese green building standards. *Sustainable Development. Special Issue: Sustainable architecture, design and housing, 20*(3), 211–221

GOBAS-Group. 2003. *Green olympic building assessment system: Architecture and building.* Beijing, China: Press

Hakkinen, T., and Belloni, K. (2011). Barriers and drivers for sustainable buildings. *Building Research and Information, 39*(3) 239–255

HM Government. (2008). Energy Act 2008. HM Government Business, Enterprise and Regulatory Reform (BERR)

HM Government. (2016). Conservation of Fuel and Power: Approved Document L. Crown Copyright. Retrieved from www.gov.uk/government/uploads/system/uploads/attachment_data/file/516016/BR_PDF_AD__L1A__2013_with_2016_amendments.pdf (Accessed 23 July 2016)

IPF. (2009). Demand for sustainable offices in the UK. Investment property forum report. Commissioned by the IPF Research Programme 2006–2009

Kasai, N., and Jabbour, C. (2014). Barriers to green buildings at two Brazilian Engineering Schools. *International Journal of Sustainable Built Environment, 3*(1), 87–95

Keeping, M. (2000). *What about demand? do investors want "sustainable buildings?"* London: RICS

Klein, A., Merkel, E., Pfluger, B., Held, A., Ragwitz, M., Resch, G., and Busch, S. (2010). Evaluation of different feed-in tariff design options—Best practice paper for the International feed-in cooperation. Technical report, Energy Economics Group, and Fraunhofer ISI

Kulczycka, J., and Smol, M. (2016). Environmentally friendly pathways for the evaluation of investment projects using life cycle assessment (LCA) and life cycle cost analysis (LCCA). *Clean Technologies and Environmental Policy, 18*(3), 829–842

Lange, R-J. (2012). Brownian motion and multidimensional decision making. Ph.D. thesis, University of Cambridge, Cambridge, UK

LCPE. (2016). London Centre of Excellence Sustainable Procurement Project. Retrieved from http://library.sps-consultancy.co.uk/documents/guidance-policy-and-practice/sustainable-procurement-toolkit.pdf (Accessed 9 July 2016)

Li, C. (2010). Regional characteristics on green building and the development of assessment system in China. *Advanced Materials Research, 113*, 598–601

Mahalingam, A., and Reiner, D. (2016). Energy subsidies at times of economic crisis: A comparative study and scenario analysis of Italy and Spain. EPRG Working Paper 1603,

Cambridge Working Paper in Economic 1608. University of Cambridge Energy Policy Research Group

Mendonça, M., Jacobs, D., and Sovacool, B. (2009). *Powering the green economy, the feed-in tariff handbook*. London: Earthscan

MEP. (2007). Ministry of Environmental Protection of the People's Republic of China. HJ/T 351–2007 National Industrial Standard for Environment Protection of the People's Republic of China: Technical Requirement for Environmental Labeling Products Eco-Housing. Beijing, China: Environmental Science Press

MOHURD and GAQSIQ. (2006). Ministry of Housing and Urban–Rural Department of the People's Republic of China (MOHURD), General Administration of Quality Supervisor Inspection and Quarantine of the People's Republic of China (GAQSIQ). 2006. GB/T 50378–2006 National Standard of China: Evaluation Standard for Green Buildings. China Architecture and Building Press: Beijing. GB/T 50378–2006

Mullich, J. (2013) Are there red flags in green construction. *Wall Street Journal*. Retrieved from http://online.wsj.com/ad/article/riskmanagement-redflags (Accessed 1 July 2016)

Nie, MS., Qin, YG., Jiang, Y., Zhang, QF. and Cai, F. (2007). *China's Eco-house technical evaluation handbook (fourth edition)*. Beijing, China: China Building Industry Press

OJEU. (2016). The Official Journal of the European Union. Retrieved from www.ojeu.com/ (Accessed 8 July 2016)

PCR. (2015). The Public Contracts Regulations 2015. Retrieved from www.legislation.gov.uk/uksi/2015/102/contents/made (Accessed 8 July 2016)

RIBA. (2011). *Green Overlay to the RIBA Plan of Work*. London: RIBA Publishing

RIBA. (2013). The Plan of Works. Retrieved from www.architecture.com/RIBA/Professionalsupport/RIBAPlanofWork2013.aspx (Accessed 1 July 2016)

Samari, M., Godrati, N., Esmaeilifar, R., Parnaz O., and Wira Mohd Shafiei, M. (2013). The investigation of the barriers in developing green building in Malaysia. *Modern Applied Science, 7*(2)

SCI. (2011). Financing and Contracting Sustainable Construction – Innovative Approaches. Sustainable Construction and Innovation Network. Retrieved from www.sci-network.eu/fileadmin/templates/sci-network/files/Resource_Centre/Reports/Financing_and_Contracting_Preliminary_Report.pdf (Accessed 7 March 2016)

Sourani, A., and Sohail, M. (2011). Barriers to addressing sustainable construction in public procurement strategies. *Proceedings of the Institution of Civil Engineers: Engineering Sustainability, 164*(4), 229–237

Sturgis Carbon Profiling. (2016). Homepage. Retrieved from http://sturgiscarbonprofiling.com (Accessed 1 November 2016)

Sun, M., Geelhoed, E., Caleb-Solly, P., and Morrell, A. (2015). Knowledge and attitudes of small builders toward sustainable homes in the UK. *Journal of Green Building, 10*(2), 215–233

TFT. (2016). Energy Survey 2016. Retrieved from www.tftconsultants.com/files/TFT-Energy-Survey-2016-Digital.pdf (Accessed 4 July 2016)

UNEP. (2009). Buildings and Climate Change Summary for Decision-Makers. United Nations Environment Programme Sustainable Buildings and Climate Initiative (UNEP SBCI)

Wiencke, A. (2012). Willingness to pay for green buildings. *Journal of Sustainable Real Estate, 5*(1), 111–133

5 Risk

The incentives and barriers for investment in sustainable buildings have been discussed in Chapters 3 and 4 respectively and this chapter is focused on risk. As previously discussed there is a perceived risk in financing sustainable buildings. Risk that the costs will outweigh the benefits, that the operational costs will be higher than standard equipment and/or that the new energy efficiency measure (EEM) will not work. These and other perceived risks are noted by the work of Durmus-Pedini and Ashuri (2010), Häkkinen and Belloni (2011), Menassa (2011), Koppenjan (2015), Bhattacharya, Oppenheim, and Stern (2015), Umamaheswaran and Jajiv (2015), Jones (2015) and Hwang, Zhao, See, and Zhong (2015). It should be noted that this research points to primarily two distinct types of risk in the creation of a sustainable buildings: (1) financial and (2) technical. Other evidence suggests that there are additional risks and the greatest risk of all is to not create a sustainable building in the first place, UNEP (2010). The body of evidence suggests that most consider financial and technical risk to be the main inhibitors to the creation of a sustainable building.

Financial risk can be reduced by examining the business case and considering options on a lifecycle basis. The business case must be thorough and credible. Investors and lenders act in a fiduciary capacity and all must be satisfied that an investment is sound, well considered and as risk free as possible. A well-developed business case is therefore essential to the success or failure of a sustainable building.

Technical risk can be reduced by understanding the technical processes involved and the type of new technology under consideration. Another approach is to pass on the technical risk to a third party and let others with experience in this area deal with any problems that may arise.

Common risk factors

With regards to sustainable buildings, all too often the decision to change a process or retrofit a solution is made late in the process and as such a full financial and technical evaluation is not undertaken correctly. The business case for a sustainable building, as succinctly outlined by Muldavin and Lowe (2006), should include a thorough description of the:

- project;
- intended users or tenants of the project;
- design and value engineering process that defined the sustainable features of the building or project;
- incentives or inducements for sustainability features that the project will benefit from or rely upon;
- project finance – detailed cash-flow projections over an appropriate duration, including projected rates of return;
- risk factors – arising from sustainable attributes or from other factors or features of the investment;
- value/cost comparison – is the project 'worth' (at completion) what it costs to deliver?

If sustainable design and sustainability thinking is integrated into the initial design process then all of these factors are considered from the onset and the building becomes a success. For example, Arup have been selected to develop HAUT, the highest wooden residential building in the world, Arup (2016). The 73-metre high residential tower located in the Amstelkwartier will include 55 apartments, public plinth hortus bicycles and an underground car park. It will have a total gross floor area of approximately 14,500 square metres and is to receive the BREEAM outstanding label. Building in wood is one of the most talked about innovations in sustainable construction internationally, due to the large storage capacity of CO_2. Using wood provides an answer to the municipality of Amsterdam's quest for CO_2 neutrality. HAUT's wood can store over three million kilograms of CO_2. In addition, 1,250 square metres of PV (solar) panels will help the building produce renewable energy, while waste water is purified through a constructed wetland on the roof. The parking garage in the building has space for electric (shareable) cars. Of course, there is a risk that these systems will not work and as a result the company may run into financial difficulty with this and any other project. This is why risk needs to be carefully considered.

5.1 Financial risk

There are a number of potential financial risks associated with the creation of a sustainable building; but in reality there is risk to consider in any project and this situation is not exclusive to a sustainable building. The main points to consider are as follows:

- Return on Investment (ROI) needs to be carefully considered.
- Inexperienced teams might lack the skills to properly implement green oriented technology which could hinder its effectiveness.
- Company budgets are not usually structured to track Life Cycle Cost (LCC) for a project making longer-term gains harder to record.
- Potential costs associated with litigation between the architect/engineers and the owner if the building is not completed.

- New green building materials might result in issues never encountered previously and be a source of litigation.
- Loss of possible financial gain if the building doesn't perform as it was intended to.
- Possible unforeseen conditions of retrofitting existing buildings.

With these risks in mind there are however a number of strategies to help reduce the risk and ensure that a sustainable building can be created. They include:

- sharing some of the financial risks with other parties by contractual agreement, for example using an ESCO;
- shifting the risk with insurance policies, for example indemnifying against possible failure;
- using tested technologies and systems to avoid future litigation, for example using EEMs that have been used on a similar building;
- restructuring the financial structure of the company to obtain LCCs in order to acknowledge the savings over the years, for example considering long-term ROI;
- seeking support for possible litigation areas, for example performance, new materials;
- commissioning and periodic re-commissioning in order to reach the best performance during the lifecycle of the building, for example implementing a Monitoring, Reporting and Verification (MRV) plan;
- retaining green building experienced team members and consultants, for example working with experienced teams;
- engaging integrated project delivery methods, for example working in a multidisciplinary fashion.

Considering all of the points above allows for decision makers to determine the cost and environmental impact as well as the building performance of green standards during design phase, allow tracking and reducing operation and maintenance cost during the occupancy phase. In any project there is a careful balance of risk vs. reward and depending upon the individuals concerned, the level of risk taken will vary considerably.

5.1.1 Risk vs. reward

When considering risk vs. reward, it is common to consider value for money (VfM) and Cost Benefit Analysis (CBA). VfM is a concept that typically considers three important aspects of economy, efficiency and effectiveness. CBA is a technique used to compare the total costs of a programme/project with its benefits, using a common metric (most commonly monetary units). This enables the calculation of the net cost or benefit associated with the programme.

Both VfM and CBA should and are generally always carefully considered when a decision is made to procure a piece of equipment for a building. The same applies to new EEMs to create a sustainable building. There is clearly an initial investment cost and over time the equipment/measure will depreciate and become less efficient, for example a gas-fired boiler. For the majority of existing building services technologies such as conventional heating and cooling systems, the investment and efficiencies are well known and furthermore the rate of depreciation and ROI is relatively predictable, further details pertaining to cost optimisation calculations can be seen in Ó'Riain and Harrison (2016). For new EEMs, this is less predictable and therefore there is an inherent higher risk. It is this risk that acts as a barrier to uptake. However, this risk can be significantly reduced by involving a third party ESCO who become the liable party through an EPC. Should you decide to work with an external third party they will also carry the technical risk.

5.2 Technical risk

In addition to financial risk, there is of course technical risk. This specifically relates to the risk of new equipment failing by not providing the necessary power, having intermittent faults and/or causing a secondary unforeseen problem. Some specific examples are provided below:

- Making a building airtight necessitates the requirement for a mechanically ventilated system to pump fresh air around the building.
- The owner of a commercial premise decides to replace the existing heating system with a new renewable EEM. If this new measure fails or is not compatible with the building it is left without heat and the occupants and not able to use the building in the winter. The repair work and/or loss in rental income may outweigh any cost savings made.
- Green roofs, which are costlier than conventional roofs, require careful design, construction and monitoring post construction.
- Improved insulation can lead to increased moisture which needs to be removed.
- New green construction materials require field testing prior to installation.
- New green construction materials require specialist installation from trained personnel. General contractors may rely on sub-contractors who are not trained correctly.

In order to reduce risk, it is common to use a risk register and a risk assessment template, see Table 5.1, whereby the risks associated with the various measures are considered. For example, the procurement and installation of a new EEM will carry a business impact risk which could be extreme or insignificant. If it is the former the option may be rejected unless there is supporting evidence to lower the risk. Following the use of a risk assessment template, a decision can then be made to proceed or reject the idea.

Technical risk exists but can be reduced through careful planning and the use of risk registers and templates. One area of concern is that of ongoing performance

Table 5.1 Risk assessment template

		Business impact				
		Extreme	*Major*	*Moderate*	*Minor*	*Insignificant*
		Complete operational failure	*Severe loss of operation, highly damaging*	*Substantial loss of operation*	*Noticeable but limited impact*	*Minimal with negligible costs*
Almost certain	Highly likely for this to happen	100%	80%	60%	25%	1%
Probable	Likely to happen	80%	60%	50%	20%	1%
Possible	Distinctly possible	60%	50%	40%	15%	1%
Unlikely	Uncommon but may be experienced	25%	20%	15%	5%	0%
Rare	Probably never occur	1%	1%	1%	0%	0%

and the fact that the predicted and actual performance of systems can and does differ. This is commonly referred as the performance gap. The type of funding in place to create a sustainable building, i.e. traditional self-funding vs. new methods, has a major impact on the level of risk inherited. Self-funding carries the financial and technical risk. Involvement of a third party i.e. an ESCO, ensures that they carry the financial and technical risk. If the second option is chosen, then the performance of the equipment and the performance gap may or may not be of concern. It would be of concern through a shared saving model and of no concern through a guaranteed saving model. Both of which are explained in Chapter 2.

5.2.1 Performance gap

The possibility of not having the building performing as it was intended is an area of great concern and a significant risk in the creation of a sustainable building. Ashuri (2010) found that green buildings perform better than non-green buildings; however, the same body of research also found that some of the buildings perform worse than the national average. The full reasons are unclear and are subject to further investigation, however information such as this might be enough to confuse decision makers about the actual real world performance of green buildings. Menezes *et al.* (2012) also commented on the significant evidence that suggests that buildings are not performing as well as expected. This statement is also supported by the work of Demanuele, Tweddell, and Davies (2010), Bordass, Cohen, Standeven, and Leaman (2001) Bordass, Cohen, and Field (2004) and Wilde (2014). So, why is this the case? Many of the factors that contribute relate to occupancy behaviour and facilities management. This is associated with the lack of feedback to designers once a building has been constructed and occupied. This has opened up a new area of investigation entitled Post-Occupancy Evaluation (POE). The Post-Occupancy Review of Buildings and their Engineering (PROBE) project investigated the performance of 23 buildings that were previously featured as 'exemplar designs' in the Building Services Journal. The studies found that actual energy consumption was typically twice as much as predicted, Bordass *et al.* (2001). In short, one can create the most sustainable building possible but if the occupants do not act as predicted the demands for energy can increase significantly. Examples include manually changing the temperature in temperature controlled rooms or poorly managed Building Management Systems (BMS). In order to avoid these problems, it is therefore necessary for all stakeholders and professionals involved – the owner, architects, engineers, constructor, developer, subcontractors and facility manager – to work in an integrated manner from the very beginning of the project through to POE, Horman et.al. (2006), Molenaar et.al. (2009). One such method to achieve this integration (in the design stage) is via the use of Building Information Modelling (BIM), which is proven to be a useful tool in identifying problematic parts of design and construction. Initiatives such as BIM allow for the determination of the cost and environmental impact of green standards during the design phase and for the tracking and reduction in operation and maintenance costs during the occupancy phase. Involving facility managers in every step of the process would add valuable input in the lifecycle and POE of the project. All of the above involves a change to existing processes and operations.

5.2.2 Process and operational change

The routes to the creation of a conventional and sustainable building have been discussed in Chapter 1 and as discussed above a fundamental shift in convention is required in order to create a viable long-term sustainable building. This process change could be via the use of BIM, a new framework for design and construction and/or a lifecycle approach. There have been a number of attempts to create a new process and/or framework for the creation of a sustainable building including for example the RIBA Green Overlap, the BREEAM methodology, the Green Building Alliance Green Building Methods, GBA (2016). Of course there is an inherent risk in being the first to use a new approach or to refurbish using a new method but those that have taken that first step and have a team of specialists to assist, are finding that the transition is not as difficult as originally envisioned and the results are better than expected. Clearly there is risk associated with process change and a useful overview of the most important risk exposures has been provided in an article written by Devine (2013) who found that:

- Initial costs for sustainable buildings are higher and as a result finding and keeping to budget the work of subcontractors is a risk carried by the main contractor.

- The availability of specialist sustainable construction materials is another risk and limiting factor. Setting up a new supplier and changing an existing process is a challenge.
- Liability exposure due to the use of new materials and/or equipment is a further risk.
- The unknown long-term performance of materials and designs is also cause for concern. Sustainable building products and materials are relatively new and as such the long term performance has not been monitored. Estimated costs savings, replacements costs and other key financial indicators could change.
- The certification systems are arduous requiring contractors to adhere to stringent protocols. Failing to reach a particular green standard i.e. BREEAM Excellent or LEED Platinum leads to additional risk.

Changing standard operating procedures and implementing an operational change is always difficult at the start, however these are necessary steps should you wish to create a sustainable building. All of which should be considered in the design stage of a building.

5.3 Risk in sustainable design

There are a number of books published in recent years to assist in identifying and mitigating the risks associations with creating a sustainable building. For example, the contractors guide to green building construction by Glavinich (2008) investigates management, project delivery, documentation and risk reduction. The handbook of sustainable building design and engineering by Mumovic and Santamouris (2009) and the handbook of green building design and construction by Kubba (2012) examined energy, health, operational performance, BIM, materials, costs and of course design.

Without the correct regard to sustainability in the design stage, each project runs the risk of (a) creating or retrofitting a building with little by the way of innovation and sustainability (b) 'adding' sustainability features at the end of the project through retrofits and (c) creating a building which is not futureproofed any may need more expensive additions in the future. This must be avoided at all costs and is why this design stage is crucial.

Creating an integrated team during design is the key to success. Generally speaking, built environment professionals (engineers, planners, architects, designers etc.) tend to work on their individual aspects of the building ensuring that they provide what is necessary within budget and to time. Typically, a client will commission an architect or design team, then the engineers get involved. Building services and structural engineers could be next, before a building contractor would undertake the work. The contractor would then subcontract the work on various parts. This fragmented approach to new building or refurbishment does not help in creating sustainable buildings. The conventional approach to sustainable design (in the United Kingdom and United States respectively) is to consider BREEAM or LEED. A new method of working to encourage integrated design, is the use of Integrated Project Delivery (IPD). A central part of which is BIM. A useful guide to IPD has been produced by the American Institute of Architects, AIA (2007) and Jones (2014) investigated IPD for maximising design and construction considerations regarding sustainability. He found that the results are conclusive that in comparison to the present process of construction development, all participants would prefer greater levels of collaboration to review and resolve design and construction problems to find the best solution. This is required at much earlier stages of the project than currently occurs.

At the design stage of the project it is essential that the correct options are chosen to create or refurbish a building to become a sustainable building. The process should be that the building fabric and materials are considered first, followed by a review of the possible energy and services requirements and the use of low and zero carbon (LZC) technologies with a final review of the alternative options. Choosing the right options at the right stages and obtaining 'buy in' from all involved is crucial.

5.3.1 Choosing the right options

In order to create a sustainable building, one must use the fabric first approach followed by the introduction of LZC technologies and review of other options. It is therefore essential that each built environment professional involved in the process is aware of the options available. Table 1.2 in Chapter 1 provided a list of Common EEMs suitable for the domestic and non-domestic building sectors. Those highlighted in bold are the most popular options for the building fabric and LZC solutions. Table 5.2 is a reproduction of Table 1.2 with an additional column that highlights the possible risk associated with that measure. This is purely a subjective analysis undertaken by the author, based upon the three criteria of economy, efficiency and effectiveness.

As can be seen in Table 5.2 a particular risk category has been applied to each of the seven common EEMs. The low risk options are those that are relatively risk free, economical, effective and efficient in delivering energy and carbon savings. The medium risk options have a high initial cost, may or may not be eligible for government funding in the present day and may lead to other additional costs if not planned correctly.

Choosing the correct option/s is not straightforward when they are combined. For example, a commercial building owner decides to super-insulate the premise creating an almost airtight building by insulating all the floors, walls and ceiling and replacing all the windows with triple glazing. The companies that occupy that building grow, recruit more staff and the occupancy level rises. As a result, the building overheats and there is now a demand for cooling. Will the owner now be required to purchase a mechanical ventilation system? Will the cost of this

Table 5.2 Risk associated with the most common EEMs

Common EEMs	*Description*	*Risk associated with measure*
Fabric solutions	**Cavity wall insulation**	**– Low**
	Draught proofing	
	Energy efficient glazing	**– Medium**
	External wall insulation	**– Medium**
	High thermal performance external doors	
	Internal wall insulation	
	Loft or rafter insulation	**– Low**
	Flat roof insulation (Warm deck – Cold deck systems or Inverted flat roofs)	
	Under-floor insulation	
	Heating pipe insulation	
LZC solutions *Heating, ventilation and air-conditioning (HVAC) Lighting solutions Micro-generation technologies*	**Condensing boiler systems**	**– Low**
	Heating controls (for wet central heating system and warm air systems)	
	Under-floor heating systems	
	Heat recovery systems	
	Mechanical ventilation systems (predominately nondomestic use)	
	Flue gas heat recovery devices	
	High efficiency replacement warm-air units	
	Fan- assisted replacement storage heaters	
	LED lighting	**– Low**
	Effective lighting controls (control gear: ballasts)	
	Ground and air source heat pumps	
	Solar thermal	
	Solar photovoltaics (PV)	**– Medium**
	Biomass boilers	
	Micro-CHP	
	Micro-wind generation	
	Micro-hydro systems	

outweigh any savings made? The answers to questions like this need to be carefully considered from the outset with risks and possible remediation measures outlined at the design stage. This is why it is important to make the correct investment decision based on the correct information.

5.4 Making investment decisions

As previously discussed, making an investment decision is based upon two key factors which are finance and risk. Finance and in particular profitability is typically measured by the ROI, however risk is not as straightforward and can be subjective. There are very few simplistic methods that built environment professionals can use to help make an informed decision. The orthodox theory is to calculate the present value of the expected profits, then calculate the present value of the expenditures and determine whether the difference between the two, the Net Present Value (NPV) of the investment, is greater than zero. If it is then this becomes a good investment. The basic rule on making investment decisions is therefore relatively simple. Calculate the NPV of an investment and see whether it is positive. A number of key questions remain: How should the expected profits be estimated? What about inflation? What discount rates should apply? How should depreciation and taxes be considered? Hammond, Keeney, and Raiffa (2015) published a practical guide to making better decisions using the PrOACT approach. They found that even the most complex decision can be analysed and resolved by considering eight elements. The first five are Problem, Objective, Alternatives, Consequences and Tradeoffs (PrOACT). The three remaining elements are Uncertainty, Risk Tolerance and Linked Decisions. To resolve any situation, they suggest that you consider each of these elements and think systematically about each, focusing on those that are key to your particular situation. Based on this approach and others, the author of this book has developed a new simpler approach to help in making investment decisions. The Energy Efficiency Value Matrix (EEVM), developed by Finnegan, dos-Santos, Chow, Yan, and Moncaster (2015), is based on the fundamentals of Value Management (VM) and is concerned with improving and sustaining a desirable balance between the wants and needs of building owners and the resources needed to satisfy them. Every building owner will want to reduce energy spend and there is a constant need to deliver the best value for all concerned. VM is based on the principles of defining and adding measurable value, focusing on objectives and function. VM provides a value-focused management style to the selection of EEMs, Designing Buildings (2013). The EEVM considers cost, flexibility, maintenance, energy saving and product sourcing. Under these headings, a user may consider any EEM. For example, the EEVM can be used to decide upon the best measures for reducing electricity use or the best ESCO to use. The building owner will need to initially consider the weighting factor that needs to be applied to each functional driver, and then decide how to score each option. Finally, scores are calculated to provide an optimal solution. In making these investment choices, the weighting applied is as a result of decision to invest in the short or long term. For example, if the building owner was seeking a short-term ROI, then that particular EEM would score poorly if the payback was greater than 10 years. Conversely, if the building owner was happy with a longer-term ROI then this particular EEM would score highly.

5.5 Guiding principles

The guiding principles for the risks to investment in sustainable buildings are listed below.

Principle 5.1 – Financial risk can be eliminated through the involvement of a third party ESCO and the introduction of an EPC.

Principle 5.2 – Technical risk can be reduced significantly by careful planning, use of established techniques and a willingness to change processes.

Principle 5.3 – Sustainable buildings should be considered in the design stage and choosing the right combination of energy saving options is crucial to success.

Principle 5.4 – Making investment decisions is subjective and dependent upon your personal circumstances. They must be considered on a long-term basis.

References

AIA. (2007). Integrated Project Delivery: A Guide. The American Institute of Architects. Retrieved from www.aia.org/aiaucmp/groups/aia/documents/pdf/aiab083423.pdf (Accessed 9 August 2016)

Arup. (2016). HAUT. Retrieved from www.arup.com/projects/haut (Accessed 9 November 2016)

Bhattacharya, A., Oppenheim, J., and Stern, N. (2015). Driving sustainable development through better infrastructure: Key elements of a transformation program. *Brookings Global Working Paper Series*, Global Economy and Development. Working Paper 91

Bordass, B., Cohen, R., Standeven, M., and Leaman, A. (2001). Assessing building performance in use 3: Energy performance of probe buildings. *Building Research & Information, 29*(2), 114–128

Bordass, B., Cohen, R., and Field, J. (2004). Energy performance of non-domestic buildings – closing the credibility gap. In: *International Conference on Improving Energy Efficiency in Commercial Buildings*. Frankfurt, Germany

Demanuele, C., Tweddell, T., and Davies, M. (2010). Bridging the gap between predicted and actual energy performance in schools. World Renewable Energy Congress XI. 25–30 September, Abu Dhabi, UAE

Designing Buildings (2013) Value management in building design and construction. Retrieved from www.designingbuildings.co.uk/wiki/Value_management_in_building_design_and_construction (Accessed 9 August 2016)

Devine. (2013). Managing risks for sustainable buildings. Retrieved from www.forconstructionpros.com/article/11222946/managing-risks-for-sustainable-building-contractors (Accessed 9 August 2016)

Durmus-Pedini, A., and Ashuri, B. (2010). An overview of the benefits and risk factors of going green in existing buildings. *International Journal of Facility Management, 1*(1)

Finnegan, S., dos-Santos, J., Chow, D., Yan, Q., and Moncaster, A. (2015). Financing energy efficiency measures in buildings – a new method of appraisal. *International Journal of Sustainable Building Technology and Urban Development, 6*(2), 62–70

GBA. (2016). Green Building Alliance Green Building Methods. Retrieved from www.go-gba.org/resources/green-building-methods/ (Accessed 9 August 2016)

Glavinich, T. (2008). *Contractors Guide to Green Building Construction.* John Wiley and Sons

Hakkinen, T., and Belloni, K. (2011). Barriers and drivers for sustainable buildings. *Building Research and Information, 39*(3), 239–255, DOI: 10.1080/09613218.2011.561948

Hammond, J., Keeney, R., and Raiffa, H. (2015). *Smart choices: A practical guide to making better decisions.* Boston, MA: Harvard Business School Press

Horman, M., Riley, D., Lapinski, A., Korkmaz, S., Pulaski, M., Magent, C., . . . Dahl, P. (2006). Delivering green buildings: Process improvements for sustainable construction. *Journal of Green Building, 1*(1), 123–140

Hwang, B., Zhao, X., See, Y., and Zhong, Y. (2015). Addressing risks in green retrofit projects: The case of Singapore. *Project Management Journal, 46*(4), 76–89

Jones, B. (2014). Integrated Project Delivery (IPD) for maximizing design and construction considerations regarding sustainability: The 2nd international conference on sustainable civil engineering structures and construction materials. *Procedia Engineering, 95*, 528–538

Jones, A. (2015). Perceived barriers and policy solutions in clean energy infrastructure investment. *Journal of Cleaner Production, 104*, 297–304

Koppenjan, J. (2015). Public–private partnerships for green infrastructures: Tensions and challenges. *Current Opinion in Environmental Sustainability, 12*, 30–34

Kubba, S. (2012). *Handbook of green building design and construction.* Oxford: Elsevier

Menassa, C. (2011). Evaluating sustainable retrofits in existing buildings under uncertainty. *Energy and Buildings, 43*(12), 3576–3583

Menezes, A., Cripps, A., Bouchlaghem, D., and Buswell, R. (2012). Predicted vs. actual energy performance of non-domestic buildings: using post-occupancy evaluation data to reduce the performance gap. *Applied Energy, 97*, 355–364

Muldavin, S., and Lowe, T. (2006). Documenting green building value: The appraisal and underwriting process. Session 204, Greenbuild 2006, Green Building Finance Consortium, San Rafael, California. Retrieved from www.greenbuildingfc.com

Mumovic, D., and Santamouris, M. (2009). *A handbook of sustainable building design and engineering.* Abingdon, UK: Earthscan

Molenaar, K., Sobin, N., Gransberg, D., McCuen, T., Korkmaz, S., and Horman, M, (2009). Sustainable high performance and Project delivery methods. A state- of-practice method. Retrieved from www.dbia.org/NR/rdonlyres/AA033026–60BF-495B-9C9C-51353F744C71/0/Sep2009ReportPankowDBIA.pdf

Ó'Riain, M., and Harrison, J. (2016). Cost-optimal passive versus active nZEB. How cost-optimal calculations for retrofit may change nZEB best practice in Ireland. *Architectural Science Review, 59*(5), 358–369

Umamaheswaran, S., and Jajiv, S. (2015). Financing large scale wind and solar projects—A review of emerging experiences in the Indian context. *Renewable and Sustainable Energy Reviews, 48*, 166–177

UNEP. (2010). Green Buildings and the Finance Sector. CEO briefing: A document of the UNEP FI North American Task Force. Retrieved from www.unepifo.org

Wilde, P. (2014). The gap between predicted and measured energy performance of buildings: A framework for investigation. *Automation in Construction, 41*, 40–49

6 Case studies

Case study 1: The University of Liverpool[1]

The University of Liverpool made the decision to create a more sustainable campus and opted to source external third-party funding. They researched the market and found that as they are a public sector organisation, their best solution was to use ESCO type Salix Finance (as described in Section 2.2.3) and in particular the Salix Energy Efficient Loans Scheme (SEELS). They used this funding to pay for Combined Heat and Power (CHP) engines in their boiler house. The total project cost was £7.3 million, with £6.1 million of funding coming from Salix and £1.2 million from other funding sources. As a result, the University is saving £1.5 million and 5,730 tonnes of carbon per year.

Project overview

The University of Liverpool embarked on a capital development programme that would increase the space of the campus and therefore the energy demand. To cope with the increased demand, the University needed to find a way to provide heat and power to their expanding campus. With a £6.1 million interest-free loan from Salix, they installed two new 2 Megawatt CHP units which help to provide low carbon emission power for the campus as well as feeding heat into the district heating network. They now benefit from greatly reduced annual energy costs and carbon emissions.

The business case for energy reduction

With 2020 carbon emission targets rapidly approaching and plans to expand over the next 5 years, the University required a solution to not only reduce their energy bill to help offset the expansion of their estate, but also to decrease their carbon emissions. Following on from the success of the existing 3.4 Megawatt CHP in the energy centre, see Figure 6.1, the University decided to install CHP generators in their disused Grade II listed boiler house. With projected savings of more than £22 million over the lifetime of the technology alongside substantial carbon savings, there was a clear business case for the installation of CHP units.

Figure 6.1 University of Liverpool Energy Centre

The solution

This project forms part of a long-term strategic programme of funding between Salix and the University. The University first established a joint-funded Revolving Green Fund (RGF) with Salix and Higher Education Funding Council of England (HEFCE) in 2009. They have also utilised the Salix loan funding programmes for projects such as lighting upgrades and heating controls as well as this CHP installation.

Lessons learnt

Peter Birch, Engineering Services Manager at the University of Liverpool, has the following advice to those considering undertaking a CHP project:

> During the procurement process, it is important to recognise the overall benefits a bid might contain in addition to the capital cost, such as the cost of an on-going annual maintenance regime which is crucial to the success of a project with a longer term life. The contractor that was awarded the tender for this project provided the capital works as well as a fixed annual fee maintenance schedule which was an attractive proposition.

Peter also suggests that any lack of metering of heat, gas and electricity should be addressed. This ensures a greater level of understanding as to where your energy is going. Finally, in order to assess multiple loads, is it crucial to have a clear understanding of the CHP system and district heating network. If there are multiple CHPs on one network, it is important to be aware of the complications associated with this.

Projected savings

Total loan value £6,100,000
Annual £ savings £1,500,000
Annual savings tonnes of CO_2 5,730
Lifetime £ savings £22,600,000
Loan payback 4.1 years

Case study 2: Blackpool Teaching Hospital NHS Foundation Trust[2]

Blackpool Teaching Hospitals NHS Foundation Trust like the University of Liverpool is a public sector body that wished to act on energy efficiency and seek to upgrade their equipment and reduce carbon emissions. It is through a third-party ESCO (Salix Finance Ltd) that this project was successful.

Figure 6.2 Blackpool Teaching Hospital NHS Foundation Trust

Project overview

Blackpool Teaching Hospitals NHS Foundation Trust, see Figure 6.2, wanted to replace their outdated heating plant equipment with more energy efficient plate heat exchangers, providing a safer working environment and increasing patient comfort. The 100 per cent interest-free Salix funding has provided the Foundation Trust with a safe, low-risk option and the upfront capital for an energy efficient project. Once the loan is repaid it will enable the savings on energy costs to be reinvested into further energy efficiency improvements across the estate.

The business case for energy reduction

The Foundation Trust was looking for an effective way to improve the efficiency of their ageing estate and reduce their energy costs. They were also aiming to improve the working environment for staff and comfort for patients. This project is expected to save them more than £2 million in lifetime savings, but will also allow them to lower their carbon emissions in line with the organisation's carbon reduction commitments.

The solution

The funding was used to upgrade the steam distribution systems at Blackpool Victoria Hospital, see Figure 6.2. The steam calorifiers were replaced with modern energy-efficient plate heat exchangers to provide heating and hot water to crucial patient areas throughout the hospital, all of which has since increased patient comfort, while also reducing maintenance costs. The Foundation Trust is looking to work closely with Salix in the coming years to help fulfil their energy efficiency plan.

Lessons learnt

Carla Wilson, Energy Technician at the Blackpool Teaching Hospitals NHS Foundation Trust, made the following statement:

> Due to the success of this project, the Foundation Trust is now in the process of developing a three year energy efficiency project plan to utilise Salix funding. Not only will this help us achieve our carbon reduction targets, but will also relieve pressure from an already strained existing budget, which is greatly needed at this time.

Projected savings

Total loan value £476,586
Annual £ savings £103,143
Annual savings tonnes of CO_2 5,730
Lifetime £ savings £2,939,576
Loan payback 4.6 years

Case study 3: St Barts Hospital[3]

St Barts is one of the oldest hospitals in London, founded in 1123, see Figure 6.3. With new energy efficiency standards and carbon reductions targets, the hospital and NHS trust sought a new method of creating energy.

Project overview

The hospital decided to install new energy efficiency measures to meet their energy needs and achieve the carbon reductions necessary under the Climate Change Act (CCA). In addition, the trust wanted to focus on longer-term strategies for investment and change. Finally, the trust wished to ensure the hospitals financial, environmental and socially sustainability targets were also met.

The business case for energy reduction

The trust was originally looking for a more effective way to improve the efficiency of their ageing estate and reduce their energy costs. With limited budgets and investment potential, the trust decided to engage with a third-party financier. As a result, they sought the expertise of an ESCO to create an Energy Performance Contract (EPC).

The solution

Following an extensive review of options, the solution came in the form of a Combined Cooling, Heat and Power (CCHP) and the delivery of a new energy

Figure 6.3 St Barts Hospital

centre. The project was made possible through the expertise of each of the partners. The financial expertise provided by the external fund enabled the designers 'Skanska' to design a solution that was compatible. As a result, savings could be used to deliver a cost neutral solution to the organisation.

Lessons learnt

Energy and Climate Change Minister Greg Barker said:

> This is exactly the type of project which the Green Investment Bank was set up for. Technology like CCHP has the potential to make significant emissions and cost savings in energy intensive operations such as hospitals. But it requires significant financial backing, so the partnership funding between SDCL and the Green Investment Bank is something I hope will be replicated to make many similar projects a reality.

Director of Estates and Facilities for Barts Heath NHS Trust, Trevor Payne, said:

> Barts Health NHS Trust are delighted to lead the way and be the first organisation to successfully use the collaboration framework model between the NHS Confederation, SDCL and GE, with our partners Skanska. The project will deliver a Combined Cooling, Heat and Power (CCHP) solution to one of the oldest and most prestigious hospitals in the world, St Bartholomew's. The agreement is the first of its kind and is being delivered through our existing PFI contract. It is an important step for us in achieving our sustainability goals and we believe that collaboration and partnership are the key to delivering successful, sustainable healthcare for the future.

Projected savings

Total loan value €3,500,000
Annual €savings €675,000 per year
Annual savings tonnes of CO_2 2,492 tonne/year
Payback 5 years
Contract length 7 years

Case study 4: Santander Bank[4]

Santander wished to improve lighting in its 791 branches across the United Kingdom, reduce energy costs and lower its carbon footprint.

Project overview

The Bank, see Figure 6.4, realised that electricity costs are higher than they should be via the use of inefficient lighting systems across all UK branches. The Bank

Figure 6.4 Santander Bank

also acknowledged that significant capital expenditure would be required and that they didn't have the in-house expertise for the wholesale fit out and replacement.

The business case for energy reduction

Through a lighting replacement programme, Santander's energy costs could be reduced by over 60 per cent, saving more than 7,000 tonnes of carbon. With a very high initial capital expenditure required of £17.5 million, Santander sought the expertise of a third party. In this case a company called SDCL managed the project and introduced an ESCO-type arrangement with the UK Green Investment Bank.

The solution

The solution was to appoint GE lighting to deliver and maintain all lighting and maintenance of new lower energy lighting solutions in each UK branch. SDCL invested in the project through a lighting services agreement. Earning their return through performance based service charges over a 10-year contract. In this arrangement SDCL took on board the financial, construction, commissioning, operational and performance risk.

Lessons learnt

Amber Rudd, Secretary of State for Energy and Climate Change, said:

> We need research, innovation and serious private sector interest in clean energy technology to make real progress in transition to a global low carbon economy. The LED lighting retrofit project is a great example of steps that can be taken by companies to cut GHG emissions.
>
> I want a global deal in Paris that will help to ensure our long-term economic security, creating a level playing field for businesses and clearing the path for private sector investment in clean technologies.

Chris Richold, director, Santander, said:

> Santander has been keen to demonstrate leadership in environmental sustainability and finance. There are fewer better ways to do so than to conduct and finance an energy efficiency upgrade throughout our entire UK estate. We look forward to working with SDCL and GE to help replicate similar solutions for our clients in the UK and around the world.

Projected savings

Total loan value £17.5 million
Annual £ savings £3 to £5 million
Annual savings tonnes of CO_2 7,000
Payback approximately 4+ years
Contract length 10 years

Case study 5: Sentinel Housing Association (SHA)[5]

Sentinel Housing Association (SHA) is investing to upgrade its homes with a range of green technologies to reduce their environmental impact and cut residents energy bills.

Project overview

In partnership with Vital Energi and as part of the RE:THINK upgrade campaign, SHA invested £5 million with a long-term aim of cutting CO_2 emission by 15 per cent and reduce fuel poverty for the residents.

The business case for energy reduction

SHA identified solar energy as the key technology and help to achieve the cost and carbon reductions necessary. SHA invested in a solar PV array to generate savings of £45,000 in energy bills and an average annual saving of £150 per household. Funding was arranged through Vital Energi over a 25-year EPC. The solar PB array was eligible for UK Governments Feed-in-Tariff (FIT) on both the generation and export rates.

The solution

The solution was to install solar PV panels to 300 of its homes in the Basingstoke and Deane and Hart District of Hampshire as a pilot scheme. The project timeline was tight and the solar PB panels were required to be fitted and operational within 12 weeks. In total 337 homes were fitted.

Lessons learnt

Carolyn Whistlecraft, SHAs Climate Change Officer, said:

> Vital Energi's professional project management and excellent tenant liaison has led to the successful delivery of over 300 solar DePV installations to an extremely tight timescale. We are happy with the quality of service and cooperation provided to the residents throughout the project. The scheme will not only deliver long term financial savings but will also help us achieve our carbon reduction targets.

Projected savings

Total loan value £3 million
Annual £ savings £45,000
Annual savings tonnes of CO_2 1.2 tonne (carbon trust estimates)
Contract length 25 years

Case study 6: Marine Corps Base Camp, Pendleton, California[6]

The Pendleton Marine Corps (PMC) Base Camp is located on the Southern Californian coast and was established in 1942 to originally train US Marines for service in World War II. The site is spread over 125,000 acres of land and consists of numerous buildings used for infantry training and residential housing, see Figure 6.5.

Project overview

The US Department of Energy (USDOE) Federal Energy Management Programme (FEMP) requested all federal agencies to reduce energy, water and Greenhouse Gas (GHG) emissions across the US. PMC Base Camp is part of the FEMP and as a result has been requested to reduce energy use.

The business case for energy reduction

In 2009 the PMC Base Camp spent approximately $25,000,000 on electricity, natural gas and fossil fuel. With energy prices rising they decided to develop an energy-reduction plan to therefore reduce energy use. They decided to use an

Figure 6.5 Marine Corps Base Camp, Pendleton, California

ESCO Energy Savings Performance Contract (EPC) and Power Purchase Agreement (PPA) model which allows federal agencies to engage with external third parties and finance onsite renewable energy projects with no upfront capital costs through a PPA with private investors.

The solution

The solution was to investigate a range of energy saving measures from lighting replacements, Solar PV and micro wind turbines through to uprated Air Handling Units (AHU) and air conditioning. As a result, the energy team at PMC Base Camp achieved a 44 per cent reduction in energy consumption, through utility energy savings, energy education and awareness and energy efficient technologies. The base saved more than $3 million per year.

Lessons learnt

Jeff Allen, Energy Manager, PMC Base Camp, said:

> To maximise energy reduction and eliminate energy waste, all energy-consuming systems are considered in our energy plan. Lighting represents a straightforward way to reduce energy consumption. It reduces energy waste, prolongs lamp life, reduces maintenance costs and contributes to our energy reduction goal.

Mark Memmott, CEO, Daylight Technology, said:

> To have automated lighting control is the best solution because it's not personally driven. If you've dependent on someone to turn off a switch when daylight occurs, it is not likely to happen. The total cost savings can be quantified across the board, because it's proven that at a certain light level, the lights are going off. Plus, the technology is now able to monitor energy use and generate hard data that can help improve lighting efficiency.

Projected savings

Total loan value £unknown
Annual £ savings approximately £3,000,000
Contract length estimated as up to 25 years

Case study 7: Kings Cross Central Development[7]

The Kings Cross development will see 25 new office buildings, 20 new streets, 10 new public buildings, the restoration and refurbishment of 20 historic buildings and 2,000 new homes, see Figure 6.6.

Project overview

The site is being developed by the Kings Cross Central Limited Partnership which consists of Argent Kings Cross Ltd (Argent), London & Continental Railways Ltd

Figure 6.6 Kings Cross Central Development

and DHL Supply Chain. Metropolitan Kings Cross Ltd is the Energy Services Company (ESCO) that was charged with the task of delivering sustainable and affordable power and heating to the entire development.

The business case for energy reduction

The vision for Kings Cross was to create a community with a long-term future that has minimal impact on the environment. The challenge was to provide an energy efficient and sustainable site without compromising the quality of living, or the availability of supply. This goal was made even more ambitious by Argent's determination to cut carbon emissions by at least 50 percent.

The solution

Vital Energi initially won the contract to install 2,000 metres of pre-insulated pipe to connect the Combined Heat and Power (CHP) fuelled energy centre to all the new homes and buildings on the development. This energy centre, was built underneath a car park, which included a 14-storey apartment block built on top.

Lessons learnt

Clare Hebbes, Senior Project Director, Argent Energy:

> Sustainability is integral to the place being created at Kings Cross. The site-wide district heating network plays a key part in our overall energy strategy and helps to make Kings Cross one of the most sustainable developments in London. We are very proud that working with Metropolitan and Vital Energi we have been able to deliver this important piece of infrastructure and that it is continuing to play an essential role in the high environmental performance of all of the buildings at Kings Cross.

Projected savings

Total loan value £9 million
Annual £ savings approximately £3,000,000
Contract length estimated as up to 25 years

Case study 8: MediaCityUK[8]

MediaCityUK is a new media enterprise zone for the North designed to become a leading international hub for the creative and digital sectors, see Figure 6.7.

Project overview

Phase one covered an area equal to 18 football pitches and provided more than 15,000 jobs and 1,000 businesses. The £650 million project was developed and

Figure 6.7 MediaCityUK

managed by Peel Media, a division of Peel Holdings. The BBC became the catalyst for the development of MediaCityUK with the announcement of their decision to search for a new Northern Centre, as a result the BBC relocated five of its departments to the new northwest site in 2011. Others moved to the site including the University of Salford.

The business case for energy reduction

The UK Governments stringent building regulations coupled with Peel's ambitious energy commitment required a new centre that was cost-effective while providing a positive social and environmental impact. The development itself had a stringent Corporate Social Responsibility (CSR) policy and as such focused on new methods of environmental excellence and greener methods to deliver heating and power. Funding for the project was derived from Peel Land and Property Group and Legal & General Capital.

The solution

The solution came in developing a district heating network that comprises over 2,000 metres of pre-insulated underground pipe connected to a 40 Megawatt tri-generation system, which simultaneously produces heat, cooling and power through a centralised Energy Centre whereby a CHP engine generates heat and electricity simultaneously while an absorption chiller creates chilled water from the recovered waste heat which circulates in a water jacket around the CHP engine.

The simultaneous generation of energy coupled with the use of recovered energy means that the system operates with very high efficiency compared to traditional forms of energy generation.

MediaCityUK is the first development in the world to become a BREEAM-approved sustainable community and achieved the highest environmental saving rating in the world through the use of a highly efficient low carbon site based tri-generation system for the local generation of heat, cooling and electricity which will save approximately 20,000 tonnes of CO^2 per annum.

Lessons learnt

Ed Burrows, Property Director, Peel Media:

> We are working hard to make sure MediaCityUK is future-proof by incorporating the very best environmental and sustainable credentials into every aspect of the development.

Projected savings

Total project value £650 million
Annual £ savings approximately £560,000
CO_2 saving: 29 per cent in comparison to conventional power and heating

Notes

1 Information retrieved from www.salixfinance.co.uk/knowledge-share/case-studies (Accessed 2 October 2016)
2 Information retrieved from www.salixfinance.co.uk/knowledge-share/case-studies (Accessed 2 October 2016)
3 Information retrieved from http://eesi2020.eu/bestpractice/barts-health-nhs-trust-cchp-project-united-kingdom/ (Accessed 2 October 2016)
4 Information retrieved from www.greeninvestmentbank.com/news-and-insights/2015/uks-biggest-ever-led-financing-package-will-cut-santanders-energy-use-by-half/ (Accessed 2 October 2016)
5 Information retrieved from www.vitalenergi.co.uk/casestudies/sentinel-housing-association/#casestudy-gallery (Accessed 5 November 2016)
6 Information retrieved from http://ppd.soceco.uci.edu/sites/ppd.soceco.uci.edu/files/users/jsumcad/Wiley%2C%20Greg-Sp%2712%20PR%20%28Energy%20r Reduction%20Camp%20Pendleton%29.pdf (Accessed 5 December 2016)
7 Information retrieved from www.vitalenergi.co.uk/casestudies/kings-cross/ (Accessed 5 December 2016)
8 Information retrieved from www.vitalenergi.co.uk/casestudies/mediacityuk/ (Accessed 5 December 2016)

Appendix 1

Definitions of sustainable buildings by region

Terminology	*Definition*	*Source*	*Region*
Passive house	A building for which thermal comfort can be achieved solely by post-heating or post-cooling of the fresh air mass, which is required to achieve sufficient indoor air quality conditions – without the need for additional recirculation of air	Passivhaus (2016)	Originated in Germany, now international but mostly in Germany, Austria and Scandinavia
Zero emission house (ZEH)	A ZEH is a detached residential building that does not produce or release any CO_2 or other greenhouse gases to the atmosphere as a direct or indirect result of the consumption and utilisation of energy in the house or on the site	Australian Zero Emission House Project Josh's House (2016)	Australia
Zero net carbon	Powered and heated by a combination of on and off site renewable energy, using fossil fuels only as back up	One Planet Living Bioregional (2016)	International
Zero net CO_2 emissions (also zero carbon, zero net carbon)	The annual dwelling CO_2 emissions ($kgCO_2/m^2/year$) from space heating and cooling, water heating, ventilation and lighting, and those associated with appliances and cooking must be zero when calculated according to the methodology in the Standard Assessment Procedure	UK Code for Sustainable Homes (now abolished)	United Kingdom
Zero net emissions	100% reduction in base building emissions from those of a specified benchmark building	Green Star, GBCA (2016)	Australia
Nearly zero energy	A building that has a very high energy performance and the nearly zero or very low amount of energy required should be covered to a very significant extent by energy from renewable sources, including energy from renewable sources produced onsite or nearby	European Directive on Energy Performance of Buildings, EPBD (2016)	European Union

Terminology	*Definition*	*Source*	*Region*
Zero energy home	100% reduction in net operational energy use compared to the HERS Reference Home	Residential Energy Services Network, RESNET (2016)	United States
Net zero energy	A net-zero energy home is capable of producing, at minimum, an annual output of renewable energy that is equal to the total amount of its annual consumed/purchased energy from energy utilities	Residential Energy Services Network, RESNET (2016)	United States
Net zero site energy	Produces at least as much energy as it uses in a year, when accounted for at the site	Torcellini, Pless, Deru, and Crawley (2006)	National Renewable Energy Laboratory (US)
Net zero source energy	Produces at least as much energy as it uses in a year, when accounted for at the source. Source energy refers to the primary energy used to generate and deliver the energy to the site	Torcellini *et al.* (2006)	National Renewable Energy Laboratory (US)
Net zero energy emissions	A net-zero emissions building produces at least as much emissions-free renewable energy as it uses from emissions-producing energy sources	Torcellini *et al.* (2006)	National Renewable Energy Laboratory (US)
Carbon neutral	Zero net greenhouse gas emissions	The 2030 Challenge, Carbon Neutral Seattle, 2030 Districts (2016) National Carbon Offset Standard (2012)	United States Australia
Climate positive	Reduce amount of on-site CO_2 emissions to below zero, i.e. generate more renewable energy than total net greenhouse gas emissions, recycle and export more water than used and reuse, reduce and recycle more waste than is generated	Clinton Climate Initiative (2016)	International

Recreated from the Australian Sustainable Built Environment Council (ASBEC)
Source: www.asbec.asn.au/files/ASBEC_Zero_Carbon_Definitions_Final_Report_Release_Version_15112011_0.pdf

Appendix 2

Passive design considerations

(a) Building use

Building use has a significant role to play in passive design. Understanding the specific use of the proposed building will determine the type of design. Clearly a warehouse used for the storage of ambient goods or an underground carpark will not require the same control of internal temperature as an office space, childcare facility or server room. In passive design all building types should and do aim to run as cost-effectively as possible and minimise energy losses.

(b) Macro-climate

Macro-climates exist over very large geographical areas and are measured through solar radiation distribution, terrain heights, distribution of land and sea and global circulation. Macro-climates interact together very closely and form climate zones. Different climatic zones will require different specialist passive design approaches. For example, solar radiation is the main passive drive in an arid environment whereas dealing with humidity would be a primary driver in a more tropical environment. The macro-climate therefore has a major role to play in selecting the right building fabric and sustainable technology, which in turn impacts on the type of financial strategy i.e. a building in an arid environment will look to optimise solar radiation for energy generation resulting in the potential need for solar photovoltaics and/or solar thermal systems.

(c) Micro-climate

A micro-climate is a climatic measurement that may exist only for a short period of time and is heavily influenced by terrain and local atmospheric properties. Solar radiation is one of the foremost design considerations which can have a dramatic effect on the buildings heating energy demand and occupants internal comfort. Outdoor temperature is critically linked with solar radiation and incoming air masses with the average yearly temperature heavily influenced any buildings configuration, heat protection and ventilation required. Daily temperature fluctuations form the strategies for passive heating and cooling. Wind is another critical planning factor and is heavily influenced not only by local terrain but also by other buildings, obstacles and vegetation. *Passive ventilation* strategies can be built around meteorological data, incorporating the air flow around buildings (determined by prevailing winds). Precipitation is the final consideration with the effects of snow and rail, heavily influencing passive design strategy. Options such as rainwater harvesting are becoming increasingly popular and consideration and knowledge of rainfall rate, patterns and topography is crucial. Understanding micro-climate is key to the choice of solution and resultant financial strategy.

(d) Building orientation

Initial site planning is required to ensure that the building orientation is right. The orientation is essentially the compass direction in which the building faces and is measured by the azimuth angle of the surface relative to true north. Successfully positioning a building (or understanding the limitations of an existing building) will heavily influence the type of design. Maximising solar gain in the correct way is essential to any passive heating or cooling strategy. Correct orientation can often reduce a buildings' energy consumption by up to 85 per cent in comparison to a building that does not consider orientation, Hausladen (2012). Generally speaking, a building that is orientated along its east/west axis is much better suited for optimal daylighting. Optimising a buildings footprint and zoning of interior spaces through understanding the suns path and angles allows for interior space to be planned correctly. Orientation to wind is another key consideration for passive cooling and ventilation. Passive cooling uses the wind to provide cooling and ventilation removing the need for mechanically assisted methods. Passive cooling uses the force of the wind to pull air through the building and is the most common and least expensive option. Passive cooling and natural ventilation are made possible when air movement takes place through pressure differences as air will move from an area of high pressure to an area of low pressure. A phenomenon known as thermal buoyancy. Ventilation is also key and orientating a building so that the axis in which the building is longer will provide the least passive ventilation. Buildings do not always have to face directly into the wind and good cross ventilation can be achieved through the design of the internal spaces and structural elements, enabling the building to channel air throughout the different directions.

(e) Building shape

A linear building is ideal for optimising a passive solar strategy. Ideally the long side of the building should be facing south. Thinner buildings also increase the ratio of surface area to volume. This will make utilising natural ventilation for passive cooling easier. A deep floor plan building will make natural ventilation much more difficult as attracting air into the centre of the building may require additional mechanical assistance.

(f) Roof shape and angle

The primary function of a roof is to provide adequate shelter against the elements. Depending upon the location of the building and resultant macro- and micro-climate, the type and pitch of roof will differ. Steeper pitches allow for better rain or snow run off in temperate or cool climates. Flat roofs in warm climates allow for better heat absorption for night time heating. In order to stay cool in the summer, passive solar buildings rely on shading systems i.e. a roof overhang or brise soleil system. These help to control the amount of daylight intrusion and are most commonly arranged in fixed positions, some are mechanical and can intelligently track the sun path to suit the needs of the internal occupants.

(g) Volume to surface area ratio

The amount of heat loss from any building envelope is proportional to its surface area, a building surface area must be minimised to achieve exceptional energy efficiency. Passive designers use a ratio known as the 'shape factor' or area to volume 'A/V ratio'. This being a buildings surface area divided by its volume. Thus, designing a building with a compact shape and a low A/V ratio is one of the most basic methods to improve energy efficiency. Determining heat loss and gain is a critical factor in calculating the integrity of any passively designed building. The greater the surface area the more heat gain/loss; small A/V ratios imply minimum heat gain and loss.

(h) Site zoning

Making the building the correct shape can reduce energy consumption; moreover the correct orientation will again result in additional energy savings. Add to that the correct site zoning and significant energy savings can be realised.

(i) Typology

A linear building which is elongated along its east/west axis allows for greater exposure to heat gains in the winter months along its southern perimeter. A deep plan building is generally much more expensive to heat and cool with natural lighting difficult to spaces located away from the external envelope. A courtyard typology has traditionally evolved from arid climate conditions. Thick exterior walls with small windows prevent high amounts of solar heat gains; consequently the interior is kept cool. The exterior walls will absorb heat during the day and release it into the building at night. Therefore the building can remain at a relatively constant temperature without the need for mechanical assistance.

(j) Thermal

Buildings primarily lose heat through their envelope and through uncontrolled ventilation/infiltration. Ensuring an internal equilibrium prevents any extreme (high/low) temperature swings.

(k) Ventilation

Ventilation of an internal environment can be supplied by either natural or mechanical systems. In some cases, there are also mixed mode systems. Natural ventilation is driven by wind or through buoyancy factors caused by temperature differences. To encourage cross ventilation there should be vents or openable windows on opposite sides of the building without major obstructions to air flow. In climates where night time air temperatures in the summer months are significantly lower than day time temperatures, night time ventilation can be used in combination with thermal mass to provide cooling.

(l) Lighting

Artificial lighting accounts for around 50 per cent of the energy used in offices and a significant proportion of the energy used in other non-residential buildings.

(m) Thermal mass

Macro-climates exist over very large geographical areas and are measured through solar radiation distribution, terrain heights, distribution of land and sea and global circulation. Macro-climates interact together very closely and form climate zones. Different climatic zones will require different specialist passive design approaches. For example, solar radiation is the main passive drive in an arid environment whereas dealing with humidity would be a primary driver in a more tropical environment.

Index